QUANTUM BLUE

THE FREE ENERGY

ZOLTAN J. KISS

Order this book online at www.trafford.com
or email orders@trafford.com

Most Trafford titles are also available at major online book retailers.

Print information available on the last page.

ISBN: 978-1-4907-9832-5 (sc)
ISBN: 978-1-4907-9833-2 (e)

Trafford rev. 11/08/2019

www.trafford.com

North America & international
toll-free: 1 888 232 4444 (USA & Canada)
fax: 812 355 4082

Gravitation is our never-ending source of energy.
Whether we use it or not is our choice.

Foreword

This book concludes my studies, or at least, brings me close to that “Everest moment", when, after describing the “Base Camps” of my ascent towards the Holy Grail of relativity, the elementary processes, gravity, magnetism, space, time, matter and all other important subjects, I am now able to reveal and formulate the enthralling practical benefit of my research, the new way of generating clean and free energy.
My conclusions and solutions are given in this book, together with all the necessary experiments for proof, based on the findings of my previous six books:

- the first (2007), which is about the definitions of time and relativity;
- the second (2009) with the description of the balance of the elementary processes;
- the third (2011) proving the theories of the first two books by experiments;
- the fourth (2013) is about the secrets of gravity and the pyramids;
- the fifth (2015) proving the unity of space and time; and
- the sixth (2017) with the definition of matter.

Our major challenge today is to find the right source for the generation of electricity, without causing damage to the environment, and to find and use the appropriate efficient and clean technology without the climate risk.

The practical objective in front of me during my studies was always to contribute to the solution of this energy generation problem. And my research, really a new view and way of approaching physics, has finally granted the results: gravity, as the source for free energy; and this has led to the new technologies, namely, the power of the pyramids and the acceleration of the *Hydrogen* process.

Executive Summary

Gravitation is the energy gift of nature, granting us a never-ending source of clean, free energy.

Gravitation has two aspects: firstly, the sphere symmetrical expanding acceleration, that is to say, the mechanical impact, and, secondly, the quantum impact. Both are at our disposal, free and without limits. Whether we use them or not is our choice.

There are two possible ways for their utilization: the high voltage format for industrial use and the low voltage format for social communication and computer technology. The source is the same the technologies of the generation are different. The acceleration of the *Hydrogen* process is for the industrial purpose. The civil structures with the principles of the pyramid are the instruments used for communication.

The permanent acceleration of the *Hydrogen* process means permanent intensity increase. The *Hydrogen* process itself is a quasi-never-ending process and the acceleration of the *Hydrogen* process has no speed limit. The speeding up has its time flow impact. The higher the speed value is the slower is the time flow. The slowing down time flow means the increase of the intensity of the process. The increasing intensity of the *Hydrogen* process is conflicting with the quantum impact of gravitation. The higher the intensity increase is the higher is the conflict.

There are two options here for us:

Either we give room for the developing conflict, thus leaving the channel of the acceleration free for being lifted by the conflict, or we fix it and, in fact, close the conflict into the channel itself.

While being lifted up by the conflict has its specific importance as well, let us concentrate here on the step by step increase of the conflict, the result of the fixing only. This second case means the increase of the internal temperature. The conflict turns into heat. The energy generation has its two components: firstly, the constant and permanent quantum impact of the gravitation and, secondly, the *Hydrogen* process, speeding up with the permanent increase of the intensity of the electron process.

The speeding up of the Hydrogen process obviously needs a drive, but, after a while, the energy demand of the drive becomes covered by the generating energy itself. From this stage on, the only source of the energy generation is the quantum impact of gravitation – conflicting with the *Hydrogen* process.

The non-stop acceleration of the *Hydrogen* process keeps the conflict alive and gravitation alone grants the energy.

The device used in the experiment for the proof of this effect was a circular channel with 14 electromagnetic quantum drives all around. The quantum drives, supplied from accumulators of 36 V, speeded up the *Hydrogen* process, 1 litre and 5 bar inside the channel. The developing conflict has eased the weight of the channel. The total mass of the channel was 11 kg. The values of the lightening of the device with the *Hydrogen* process inside were around 10-60 gram! These results were long discussed with the manufacturer of the balance, a world-respected firm, and finally agreed proven. It was not any failure of the balance. It was repeated more than 200 times, proving the easing of the weight in each case.

The drives of the acceleration were electromagnets with specific wiring. One end is drawing (attracting) but the other blows (repels) the *Hydrogen* process ahead. The development of the quantum drive needed a clear understanding of magnetism. The longest section in this book is devoted to the definition and the explanation of magnetism.

Magnetism is the quantum impacts of the anti-electron processes. Magnetism is the natural component of the elementary processes. The anti-electron processes are the "consciences" of the elementary processes. They keep the processes on their own. The anti-processes do not allow the change of the elementary processes in natural circumstances. Gravitation is the anti-electron process quantum impact of the elementary processes of the *Earth*.

The anti-electron processes protect the identity of the elementary processes, but there is a limit. There are conflicts, which cannot be resolved. These conflicts are with such extremely high temperature increase and heat generation that the only way to resolve it is the step back on the periodic table. The limits of the temperature increase and the heat generation are always one and the same, but the higher the periodic numbers are, the more are the reserves for working it out. The transmutation is the move in the opposite direction entirely to the plasma if it should be necessary.

There is only one elementary process which can withstand all external impacts. This is *Hydrogen*.

The pyramids are our legacies with their still acting quantum impacts. They are establishing a system for their own communication on the surface of the *Earth*. Losing energy, as part of this system, means the accommodation of the others to the new circumstances. The energy source of the pyramids is the quantum impact of gravitation, approaching the basic surface from below. The voltage is low; the amperage is divisible as it is needed for us. And what do we need more today for the technologies used in our social and business communication than an electrical current with low level voltage and even lower amperages?

The quantum impact of gravitation for the pyramids is limitless, as long as they stand on the *Earth's* surface. When lifted up from the *Earth's* surface, they lose their internal conflict and become useless for the energy supply, as was proven by the experiments at small scale.

The world is bipolar. The category "large" certainly necessitates the existence of the “little”, in the same way as the "dark" certainly needs the existence of the "bright" and so on. If the mechanical impact can generate light, then the light impulses will also be able to generate mechanical impacts. Matter is information, with the widest scale of intensities and frequencies. This is, in fact, the reason why their appearances are so very different. The mechanical impact of the light impulse with frequency easily fits into our everyday practice. The detection of the impact needs specific preparations with the tools of the required sensitivity.

Given in this book are the addresses in "YouTube" for the demonstration, by experimentation, of all the theories discussed.

The previous books with the explanations of the subjects in detail are:
The Energy Balance of Relativity (2007);
The Quantum Energy and Mass Balance (2009);
Quantum Engine (2011);Gravitation: our quantum treasure (2013)
The Quantum Impulse and the Space-Time Matrix (2015)
Matter – the matrix of Information (2017)
- all published by Trafford Publishing Bloomington.

Table of Content

1
Matter is the matrix of information

Processes exist in processes.

The *electron* and *anti-electron processes* are the drives of the elementary communication and conflicts, establishing balance and harmony; the drives and the safeguards of the identity of the appearance. The *plasma* is the beginning, the start of the elementary evolution, the infinite *high* intensity; the *Hydrogen* process is the last one, the elementary process with infinite *low* intensity. Between these two ends are all other elementary processes, either with electron process surplus or deficit. The surplus is the drive of the elementary communication; the deficit means strong elementary structure.

The elementary processes communicate, resolving existing conflict/s, establishing the status of the balance, even modifying the aggregate status. No communication – in fact, means no matter.

The *proton process* is *sphere symmetrical expanding acceleration*; the source of energy; the step by step drop of the intensity, the internal energy capacity of the elementary process; accumulating on the anti-side; the one, continuing the *anti-proton/proton inflexion* of infinite high intensity, ($\lim dt_{xi} = 0$), the start of the expansion.

The expansion of the proton process goes with the intensity of the elementary process ε_{xp} and ends, reaching $\lim i_x = c_x$.	$\Delta e_{xp} = \frac{dmc_x^2}{dt_p \varepsilon_{xp}}\left(1-\sqrt{1-\frac{i_x^2}{c_x^2}}\right);$	1A1
This is the start of the electron process, the *sphere symmetrical expanding acceleration* with constant speed increase of $\lim(c_x - i_x)$ with the certain intensity value ε_{xe} until: $\lim \Delta e_{xe} = 0$.	$\Delta e_{xe} = \frac{dmc_x^2}{dt_i \varepsilon_{xp}}\left(1-\sqrt{1-\frac{(c_x-i_x)^2}{c_x^2}}\right);$ $\varepsilon_{xe} = \frac{\varepsilon_{xp}}{\varepsilon_{xn}};$	1A2 1A3
The *neutron process* is the *sphere symmetrical accelerating collapse*, driven by the electron process quantum drive;	$\Delta e_{xn} = \frac{dmc_x^2}{dt_n \varepsilon_{xn}}\sqrt{1-\frac{(c_x-i_x)^2}{c_x^2}}\left(1-\sqrt{1-\frac{i_x^2}{c_x^2}}\right);$	1A4

The neutron process is collapse, the increase of the intensity; the accumulation of the energy. The collapse needs external drive; the drive is the *electron process*, but the collapse has its own intensity, coefficient of ε_{xn}. The end of the collapse is the *neutron/anti-neutron inflexion*, the start of the anti-process.

The anti-proton/proton and the neutron/anti-neutron inflexions are the changes of the polarity of the elementary processes. The only events with zero time count $\Delta t_{inflexion} = 0$

The *quark* processes are the carriers *of the intensity change of the* processes, the proton and the anti-neutron expansions, the neutron and the anti-proton collapses.

Anti-processes are controlling the identity of the elementary processes. The anti-directions are the mandatory components of the elementary balance. The anti-proton/proton and the neutron/anti-neutron *inflexions* stabilise the elementary cycles, the continuity of the elementary evolution. Elementary cycles happen in infinite numbers in parallel within the elementary processes.

The *quantum impacts* are the *blue shift impacts* of the electron processes; the drives of the elementary communication, generating the balances and conflicts, establishing the aggregate status of the appearance of the elementary process.

The elementary processes are the specific phases of the elementary evolution:
- the step by step expanding acceleration in one direction, up to the overall elementary status of infinite low intensity, the *Hydrogen* process;
- the *Hydrogen* processes do not have their own elementary ends; they are accumulating; the overall collapse is unavoidable;
- the collapse is the overall *inflexion* of infinite high intensity, the *plasma* with density and temperature of infinite high values, the change of the polarity of the elementary evolution, the start of the next cycle.

The intensities of all new elementary cycles are of reduced values in each new cycle. The reduced intensity is the result of the permanent generation of the *quantum impulses,* the quantum entropy products of the elementary processes. The quantum impulses are the remains of the intensities of the electron processes, with no capacities for drive any more.

1A5 $$\lim \Delta e_{xe} = \lim \frac{dmc_x^2}{dt_i\varepsilon_{xp}}\left(1 - \sqrt{1 - \frac{(c_x - i_x)^2}{c_x^2}}\right) = 0;$$

The *quantum impulses* cannot disappear; “nothing” as such does not exist. This way they are accumulating and they set up the quantum space; increasing it continuously, building up the global frame of the elementary evolution in time and space.

The elementary evolution happens in *space* and *time*, establishing the *space* and allocating the *time* in parallel. Elementary processes happen in space with certain intensity, in time, the reciprocal value of the intensity. All processes of the elementary evolution from one end, the plasma to the other with the *Hydrogen* processes have their own specific space-times. The elementary communication happens in space and time, with certain quantum speed value of the quantum communication, different for each elementary process.

The *densities* of the elementary processes characterise the number of the elementary cycles, acting in parallel in the space-time of the elementary process.

All those, surrounding us in the nature are events, happening in time, having their own intensities. The intensity of the event is the one controlling/defining the form of the appearance. The intensities of the processes are in fact information. The *matter* means the *matrix* of the information in balance, the quantum impact of the conflicts, the speed of the quantum communication, the status of the elementary process in space and time.

1.1 The *information* = matter

S. 1.1

Once the category of "matter" is identical to the category of "information" (in global terms intensity and energy), this relation shall be valid in its opposite way as well: the information is identical to the *matter*. While the conventional definition of the *matter* is identical to the definition of the *mass*, it should not mean at all that the *mass* – as one of the measurement characteristics of the matter – would mean or would be equal to the matter. In fact, the matter/information matrix covers all existing physical impacts.

The transformation of *mass* into energy does not cause any problem in conventional terms. This direction is more or less obvious and has its different forms. The story is much more challenging in the opposite direction: the transformation of the information (intensity, energy) into mass.

The information is quantum impact. The most prevailing and classical form of the quantum impact is the *blue shift* of the electron process, but the same way all other elementary effects in fact are quantum impacts as well. The exchange of the intensity between the proton/neutron, anti-neutron/anti-proton processes are the certain forms of the quantum impacts. [The quantum impulse of the elementary processes (the quantum) is also *blue shift* impact, just of infinite low intensity, the quantum entropy status with no quantum drive potential of the elementary processes.]

The *blue shift* impacts (of the electron or anti-electron processes) are the quantum drives of the elementary processes leading to the inflexions, the starting points of the anti-sides. The sphere symmetrical expanding accelerating of the proton processes results in the electron processes, the collapse of the expanded electron process results in the neutron process. The aggregate status is established by the intensity values and the conflicts of the electron processes. The *blue shift* quantum impacts in conflict or without, either in internal or external elementary relations are the carriers of the quantum information. There is no quantum impact without information.

Elementary processes generate quantum impacts; the quantum impacts are effecting other elementary processes – communicate. Conflicts result in the change of the intensities, the conflicts generate energy. Categories intensity and time are each other's inverses. Events happen in time and this way with intensities. The events have their direct impact on the time flow.

All equations of the elementary processes are about the change of the matter in time.

All processes of the elementary evolution mean the change of the forms of the matter.

$$e_x = \frac{dmc_x^2}{dt_i\varepsilon_x}\left(1 - \sqrt{1 - \frac{(c_x - i_x)}{c_x^2}}\right); \qquad \Delta m = \frac{e_x \cdot \Delta t_i \cdot \varepsilon_x}{c_x^2 \cdot y} = \left[\frac{kg\frac{m^2}{s^2}}{\left(\frac{m}{s}\right)^2} \cdot s\right] = [kg]; \text{ indeed ;}$$

1B1 1B2

1B3 The *matter* means the *matrix* of the intensity of the process divided by the quantum drive of the process:

$$m = \int_1^{\frac{1}{\sqrt{1-\frac{(c_x-i_x)^2}{c_x^2}}}} e_x \frac{\varepsilon_x}{c_x^2}$$

The intensity of the process in global terms is the information in time.

If the definition of the *mass* in its conventional way is the attribute of the *matter*, or formulating it in other way, if the *mass* corresponds to a certain definition of the *matter*, with reference to 1B3, the matrix of the *change,* the *conflict* of the *quantum impacts* of the electron processes gives the definition of the *matter* as *information*!

In our circumstances on the *Earth* surface, the *mass* is the appearance of the conflicts of the intensities; the infinite slow gradient of the release of the intensity.

The *weight* means the conflict of the intensity of the event with the quantum impact of *gravitation*:

1B4

$$G = m \cdot g = e_{x-} \Delta t_i \frac{g}{c_x^2} = \left[kg \frac{m^2}{s^2} \frac{1}{s} s \frac{s^2}{m^2} \frac{m}{s^2}\right];$$

The mass is the appearance of the quantum information of certain density. The elementary processes differ not just in their numbers of the proton and neutron processes, their numbers and the surpluses of the anti-electron processes, but also in their numbers of the elementary cycles happening in parallel as well.

As the intensities of the quantum impacts of the anti-electron processes are equal for all elementary processes, the number and the surplus of the electron processes are the ones making the difference between the elementary processes.
The quantum impact of the anti-electron process is:

1B5

$$e_{x-} = \frac{dmc_x^2}{dt_i \varepsilon_{x-}} \left(1 - \sqrt{1 - \frac{(c_x - i_x)^2}{c_x^2}}\right) = const;$$

There is a difference between the intensities of the proton and the neutron processes. Therefore the numbers and the summarised quantum impacts of the generating anti-electron and electron process quantum impacts are different. The difference is a number, taken as n_x.

1B6 From the elementary processes it follows that $e_x = n_x e_{x-}$ and $n_x = \frac{1}{\varepsilon_x^2}$;

$$\left[n_x \frac{dmc_x^2}{dt_i \varepsilon_{x-}} \left(1 - \sqrt{1 - \frac{(c_x - i_x)^2}{c_x^2}}\right) = \frac{dmc_x^2}{dt_i \varepsilon_x} \left(1 - \sqrt{1 - \frac{(c_x - i_x)}{c_x^2}}\right) \right]$$

1.2
The intensity of the *quantum* impact

S. 1.2

The anti-electron process is the drive of the collapse of the anti-proton process. The collapse is reaching the anti-proton/proton inflexion with infinite high intensity. The inflexion is the turnaround point, the change of the polarity of the process with no duration $\Delta t_{infl} = 0$ and of infinite high intensity. This is the point, when the elementary process reaches its maximum energy capacity at maximum intensity. The capacity means the concentration and the conflict of the *quantum impacts* – which in fact: information!

The *inflexion* is the elementary status of the *end* of the *increase* of the energy potential, *the start of the release* of the energy. The inflection at the global level of the elementary evolution is the *plasma*; the matrix of infinite high intensity.

$e_{inf} = \frac{dmc_x^2}{\sqrt{1-\frac{i_x^2}{c_x^2}}};$	the final intensity of the anti-proton process; the accumulation of the energy on the anti-side, up to the *inflexion*	$e_p = \frac{dmc_x^2}{dt_p\sqrt{1-\frac{i_x^2}{c_x^2}}}\left(1-\sqrt{1-\frac{v^2}{c_x^2}}\right);$	the intensity of the proton process; the release of the intensity

1C1

The conflict of the *blue shift* quantum impacts of the electron processes of infinite high intensity results in the increase of the density of the quantum impacts and the temperature up to infinite high values. In other formulation: the temperature or the increase of the temperature means the conflict itself; the parameter of the proof of the conflict.

The matrix of this conflict is the quasi *plasma* process, where the quantum impacts are in infinite high conflict, with the densities and the temperatures of infinite high values.
The after-inflexion-plasma status is the continuous loss on its intensity and on its conflict.

The matrix of the quantum impact – the information – after the inflexion, starts the expansion for the count of its own intensity (energy), accumulated during the collapse. The expanding status is the generation of the energy, the release of the conflicts step by step. The intensities and the temperatures become less and less, the number of the quantum impacts increases the quantum space step by step.

All elementary processes go through these stages, formulating the elementary matrix, representing the different aggregate statuses of the matter.

The step by step release of the conflicts happens in different space-times, in different intensities and for different time counts. The high intensity of the post-plasma status with quantum speed of infinite high value has in fact infinite short time duration. Further expansion means the decreasing values of the intensity and the less speed values of the quantum communication. The time count, the time duration of the processes are increasing in parallel.

Conflicts with low intensities mean less with high intensities mean higher temperatures. The aggregate statuses of the processes depend on the intensities of the electron processes. The gaseous and liquid statuses mean conflicts; the solid status is without conflict.

The measured duration of the conflicts and the collapsing/expanding processes depend on the intensities of the space-times, where the measurements are taking place.
The supposed lifetime of the proton process is of quasi infinite length in our space-time. The dynamism of the change in our space-time therefore can be considered as quasi static matrix.

Ref. 1B10 1B11 of M=MI

The question is: Once the *mass* corresponds to the matrix of information, why in this case the speed limit of its acceleration is $\lim i_x = c_x$?

1C2

With reference to the earlier studies, the cyclical change of the quantum speed of the information is: $c_{x+1} = c_x\sqrt{1 - \frac{(c_x - i_x)^2}{c_x^2}}$;
(Ref. Sec. 1 of the book "Matter = Matrix of Information)

The intensity of the electron process (the drive) of the neutron collapse is:

1C3

$$e_{ex} = \frac{dmc_{x+1}^2}{dt_i\varepsilon_x}\left(1 - \sqrt{1 - \frac{(c_{x+1} - i_{x+1})^2}{c_{x+1}^2}}\right) = \frac{dmc_{x+1}^2}{dt_i\varepsilon_x} - \frac{dm}{dt_i\varepsilon_x}\left(c_{x+1}\sqrt[2]{1 - \frac{(c_{x+1} - i_{x+1})^2}{c_{x+1}^2}}\right)^2;$$

1C4

having the speed value, running out from the driving impact and establishing the start of the neutron process: $c_{x+1}\sqrt[2]{1 - \frac{(c_{x+1} - i_{x+1})^2}{c_{x+1}^2}}$;

1C5

The value of the quantum speed (the drives of the anti-electron process, driving the anti-proton process to the inflexion, where the elementary cycle starts from), is: $c_{x+2} = c_{x+1}\sqrt{1 - \frac{(c_{x+1} - i_{x+1})^2}{c_{x+1}^2}}$;

1C6

Obviously, $c_x \neq i_x$; $c_{x+1} \neq i_{x+1}$ and $c_{x+2} \neq i_{x+2}$, otherwise there would be no electron process impact any more.
At the same time it also demonstrates, there is no way that the electron process would run out from its all available intensity capacity: $\lim i_x = c_x$ is indeed!

1C7

This way, the answer is: The reason is not the supposed stage of the *mass*, rather the need of the continuity of the cycle.

1C8

The electron process cannot expand in full – there is no way to continue the drive to zero intensity: $e_{exrem} = \frac{dmc_x^2}{dt_i\varepsilon_x}\sqrt{1 - \frac{(c_x - i_x)^2}{c_x^2}} \neq 0$

The neutron process always starts from the infinite low intensity remains of the electron process. The balance of the proton/neutron processes shall be given as

1C9

$$\frac{dmc_x^2}{dt_p\varepsilon_p}\left(1 - \sqrt{1 - \frac{i_x^2}{c_x^2}}\right) = \vartheta\frac{dm}{dt_n\varepsilon_n}\left(c_x\sqrt[2]{1 - \frac{i_x^2}{c_x^2}}\right)^2\left(1 - \sqrt{1 - \frac{i_{x+1}^2}{c_{x+1}^2}}\right);$$

where $\vartheta > 1$ and $c_{x+1} = c_x\sqrt[2]{1 - \frac{i_x^2}{c_x^2}}$;

Without the coefficient ϑ the two sides of the equation above are clearly different.

There is a generating intensity capacity at the anti-proton/proton inflexion. This capacity, available at this stage cannot be fully used in the elementary process.
There are the *quantum impulses*, generating in each elementary cycle in the direct and in the anti-direct directions as well. =
= The *quantum impulse* is the process with the smallest intensity –
– the building stone of the space (the Universe):

$$\Delta_q = \frac{dmc_x^2}{dt}\left(1-\sqrt{1-\frac{i_x^2}{c_x^2}}\right) - \frac{dmc_x^2}{dt}\sqrt{1-\frac{i_x^2}{c_x^2}}\left(1-\sqrt{1-\frac{i_{x+1}^2}{c_{x+1}^2}}\right) > 0\,; \qquad \text{1C10}$$

The intensity of the quantum impulse is of infinite low value. There is no way the elementary processes could use it as drive. The only function the quantum impulse is capable for is the taking part in the quantum communication, in the handing-over of the information, generated by the elementary processes.

Δm or dm is equivalent to the quantum impact of the process; the quantum impact of the step by step collapse and expansion. The aggregate formats of the appearances of the elementary processes depend on the intensities of the conflicts of the electron process *blue shift* impacts. They have different formats: *gaseous* with high intensity, *liquid,* with increased intensity, and *solid,* with low intensity.

The all over status of the matter however still remains information, as the key of the existence is intensity.

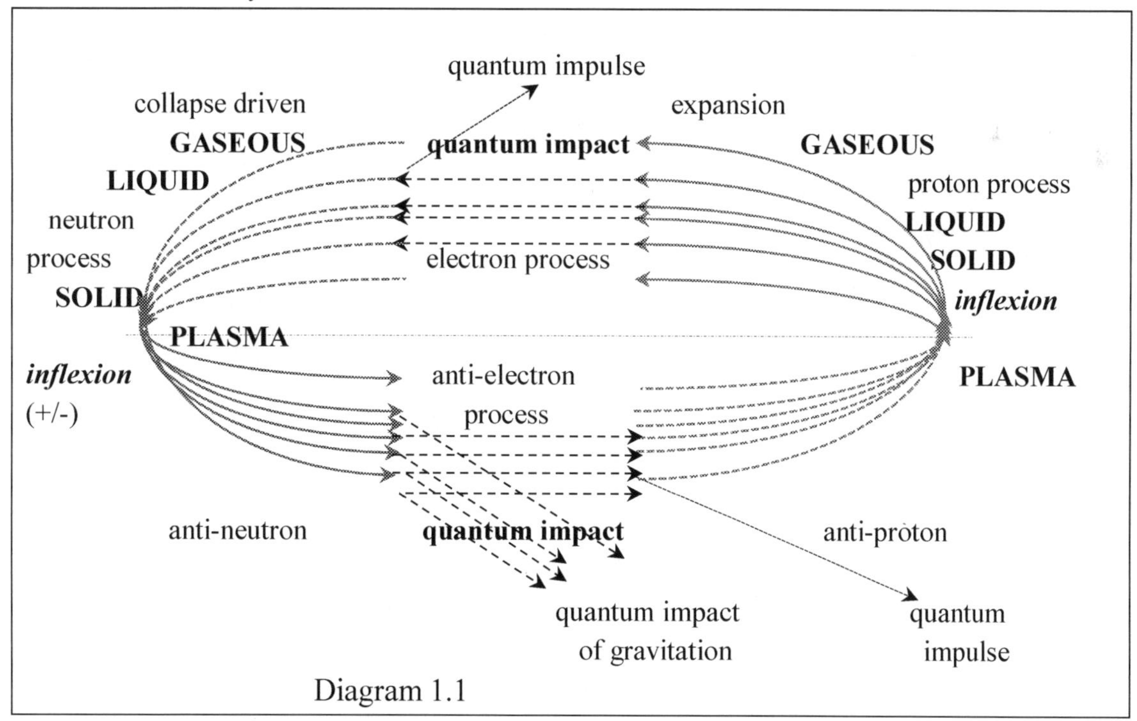

Diagram 1.1

Diag. 1.1

The electron process is the drive of the collapse of the neutron process. The inflexions of the neutron/anti-neutron processes are similar to the anti-proton/proton processes on the anti-side, driven by the anti-electron processes.

S.
1.3

1.3
The category of the *mass* and the principle of the *Hydrogen* based energy generation

The normal *Hydrogen* process of the Periodic Table has only proton and electron processes. The neutron process is missing and the anti-cycle with the anti-electron process is also missing. The *Hydrogen* process has no elementary cycle. The *Hydrogen* process itself is the last step in the elementary evolution. The inflexion of the step by step accumulating *Hydrogen* is the start of the *plasma* the new global cycle of the elementary evolution.

The acceleration of the *Hydrogen* process has its specifics. The proton process starts from the inflection of the accumulating anti-proton process of the *Helium* process.

The electron process is *blue shift* impact of infinite low intensity with infinite long lifetime:

1D1

$$e_H = \frac{dmc_H^2}{dt_i \varepsilon_H}\left(1 - \sqrt{1 - \frac{(c_H - i_H)^2}{c_H^2}}\right); \qquad \lim \varepsilon_H = \infty ,$$

The intensity of the quantum impact of the electron process e_H is of infinite low value.

1D2 The intensity, result of the acceleration is: $e_{Hv} = \frac{dmc_H^2}{dt_i \varepsilon_H \sqrt{1 - \frac{v^2}{c_E^2}}}\left(1 - \sqrt{1 - \frac{(c_H - i_H)^2}{c_H^2}}\right);$

The electron process is the final stage of the acceleration.
The proton process has its real intensity and real parameters:

1D3

$$e_{Hp} = \frac{dmc_H^2}{dt_{Hp}\sqrt{1 - \frac{v^2}{c_E^2}}}\left(1 - \sqrt{1 - \frac{v_H^2}{c_H^2}}\right); \qquad \text{or} \quad e_{Hp} = n_{Hp}\frac{dmc_H^2}{dt_{Hp}}\left(1 - \sqrt{1 - \frac{v_H^2}{c_H^2}}\right)$$

As neither the quantum drive of the electron process of the *Hydrogen* process is used, nor the generating quantum impact energy/intensity of the proton process is handed over to the neutron process, which in fact is of infinite low intensity, there is an accumulating quantum impact potential at the end of the *Hydrogen* processes, free for the communication and for the conflict with the quantum impact of gravitation.
The conflict of the quantum impacts of the *Hydrogen* process and the gravity are potential sources of the generation of the energy!

Ch.2

2
Relativity and indeterminacy

Relativity is one of the basics. Precise definition of events can only be given, if all possible variants are assessed in their appearances.
Relativity is not (just) about the time relations. Relativity is about the varieties and the forms of the appearances of the events; relativity – alongside the time count – is one of the most visible parameter of the elementary relations.

The *indispensable* condition of the matter means: events happen only once!
Any repetition would contradict to the continuity of the time flow. It would mean the end of the matter. As *Heraclitus* the philosopher has formulated it in his precisely global and famous saying: "*No man ever steps in the same river twice!*"
The *quantum entropy*, the generation of the quantum impulse guarantees the protection from the repetition of events.

The legacy of the *Greek* philosophers proves: the relativity is not the invention of the XIX century. The relativity is the principle of the dialectics; the balance and the harmony of the nature; the unity of the opposites; the direct and indirect sides of processes. There is the demand for the careful assessments of all possible options in the nature. This principle was born before our era but has still remained one of the basics of the relativity.

Zeno's paradoxes have not been resolved, as nobody has ever considered using this principle – the relativistic approach. The two most famous paradoxes are:

1. *Achilles and the tortoise*

"In a race, the quickest runner can never overtake the slowest, since the pursuer must first reach the point whence the pursued started, so that the slower must always hold a lead."

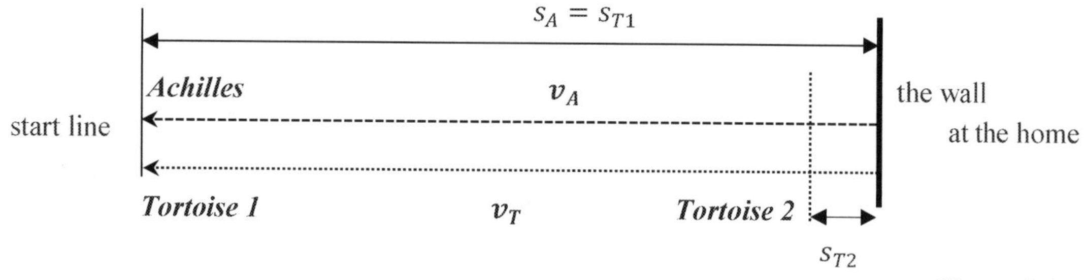

Figure 2.1

Fig. 2.1

For solving the paradox we have to use the principle of the relativistic approach. For this reason we suppose that the home is the one in motion and is the one, which is reaching the runners. The two runners stay calm at the start. This will not change the original plot of the paradox, but gives the solution.

From the point of view of the runners the home has different speed of the approach. Relative to the tortoise the speed of the home v_T is clearly less, relative to *Achilles* v_A it is much-much more.

As $v_A > v_T$ there is no question, the home reaches *Achilles* first.
The home would reach the *tortoise* first only, if the tortoise would be placed at such a distance, so close to the start place of the running home that for the same duration the home would make less distance relative to (reaching) the tortoise: $s_{T2} < s_A$.

2. *The Moving Rows*
"There are two rows of bodies, each row being composed of an equal numbers of bodies of equal size, passing each other on a race-course as they proceed with equal velocity in opposite directions, the one row originally occupying the space between the goal and the middle point of the course and the other that between the middle point and the starting-post. This involves the conclusion that half of a given time is equal to the double of that time."

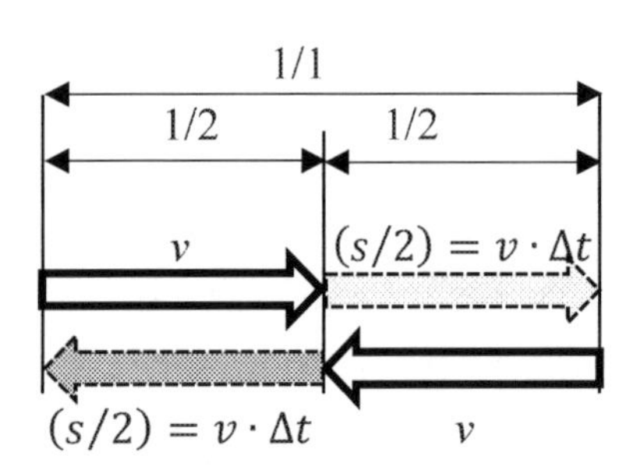

The two rows in motion with equal speed *v* in opposite directions make the full distance for Δt time period: $s = v \cdot \Delta t$;
The distance each row makes is:
$(s/2) = v \cdot \Delta t$; which for the full distance gives: $s = v \cdot 2\Delta t$

The concern is: Is Δt really equal to $2\Delta t$?

Fig. 2.2

Figure 2.2

The answer is easy, if the motion of the rows is taken in relative terms:
The relative speed of each of the rows in motion relative to the other is: *2v*.
This is the speed both of the rows are reaching the other end. The distance they together make is the full length of the path: $2(s/2) = v \cdot \Delta t = s$. (2A1)
Assessing this way there is no concern, the duration of the event is Δt. And $\Delta t = \Delta t$.

The following sections will prove, relativistic approach helps to understand the secrets of the world, to find definitions of physical phenomenon, like *gravitation* and *magnetism*.

S. 2.1

2.1
Are the numbers π and *e* irrational indeed?

In the case events are in *circulation, rotation* or in any other *cyclical* motion, each cycle shall represent a new "story" rather than the repetition of the previous one. The repetition of events contradicts to the definition of time: *time* could not be defined without change; cyclical repetition would not mean a real process, whatever the duration of the repetition was.

While the elementary processes are cyclical indeed, each of the cycles has quantum impact remains, the *entropy product* of the cycle: the *quantum impulse,* the building stone of the space. Each elementary cycle will always be different. The new cycle is never the repetition of the previous one. Either the starts or the ends are different.

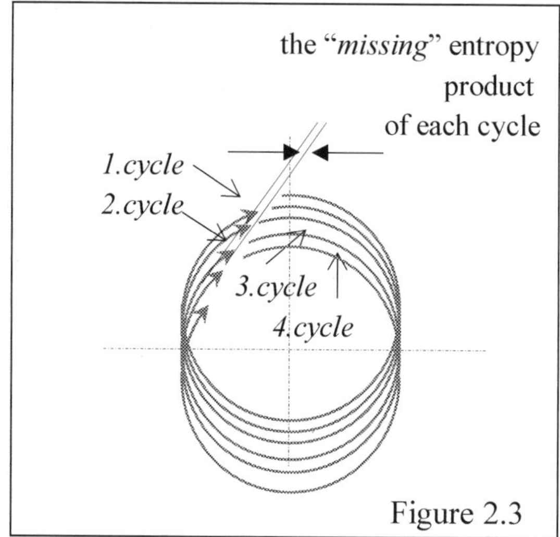

Figure 2.3

1. Each cycle starts from a new beginning, as the quantum entropy product makes the difference:
2. Each new cycle brings the starting point of the cycle forward, so the actual length of the cycle becomes always shortened.
3. The cyclical process becomes a spiral rather of repeating circle.
4. The spiral format is one of the well-known examples of the time-flow.

Fig. 2.3

The definition of the *circle* includes $\boldsymbol{\pi}$**,** a certain entropy product. Without $\boldsymbol{\pi}$ the circle would be the classical examples of the repetition of the cycle. $\boldsymbol{\pi}$ has its never ending value without the cyclical repetition of any numbers and this way guarantees the continuity of the time flow.

$\boldsymbol{\pi}$ - called as *irrational* is the natural and mathematical guarantee for avoiding repetitions.

$\pi = 3.14159265358979323846264338 32...$	π is just seen as a random variety of digits.	2B1
or $\frac{2}{\pi} = \frac{\sqrt{2}}{2} \cdot \frac{\sqrt{2+\sqrt{2}}}{2} \cdot \frac{\sqrt{2+\sqrt{2+\sqrt{2}}}}{2} \cdots$	The longer the radius is, the higher is the chance for expressing it in more precise format.	2B2
or $\frac{\pi}{2} = \frac{2}{1} \cdot \frac{2}{3} \cdot \frac{4}{3} \cdot \frac{4}{5} \cdot \frac{6}{5} \cdot \frac{6}{7} \cdot \frac{8}{7} \cdot \frac{8}{9} \cdots$	In our circumstances and space-time the 4 digits definition after the dot gives the necessary technical preciseness.	2B3
or $\pi = \frac{4}{1} - \frac{4}{3} + \frac{4}{5} - \frac{4}{7} + \frac{4}{9} - \frac{4}{11} + \frac{4}{13} - \frac{4}{15} + \cdots$		2B4

In the case of the linear motion the explanation is even more global and it is similarly valid for cyclical and for any kind of other motion as well. The answer is given by the *derangement concept*, the *inclusion-exclusion principle,* developed by P. R. de Montmort, and N. Bernoulli: "Derangement is a permutation of the elements of a set, such that no element appears in its original position. Derangement is a permutation that has no fix point."

A classical relativistic approach.

- any event, whatever the event is about, cyclical, linear or other, can be divided into infinite ($\lim n = \infty$) number of elementary process sections, elementary processes;
- the *permutation* of this number of sections and processes is: $p = n!$ factorial;

- derangement means the number of permutations, where none of the process sections occupies its original position;

2C1 in the case of process sections *n* the number of the *derangements* is $D_n = !n$ subfactorial.

(As an example: In the case the number of the process sections is 4, the number of permutations is 24 and the number of derangement – when none of the process sections appears in its original position – is 9 subfactorials.)

2C2 $$!n = n! \sum_{i=0}^{n \to \infty} \frac{(-1)^i}{i!} \; ; \quad \text{and} \quad \lim_{n\to\infty} \frac{!n}{n!} = \frac{1}{e} \; ;$$

where i – (*here*!!) is the actual number of the process section $[0! = 1]$

and - $n!$ is the number of *permutations* and

- $!n$ is the number of derangement *subfactorials.*

2C3 and this way: $$\frac{1}{e} = \sum_{i=0}^{n\to\infty} \frac{(-1)^i}{i!} \; ;$$

where the missing from the *derangement* variant, the real sequence of the process is: (as $n! = !n * S$)

2C4 $$\frac{1}{\sum_{i=0}^{n\to\infty} \frac{(-1)^i}{i!}} = S$$

The quotient of the number of the permutation and the number of the derangement in the case of $\lim n = \infty$ is

2C5 $e = \lim_{n\to\infty} \left(1 + \frac{1}{n}\right)^n$; or 2C6 $e = \sum_{n=0}^{\infty} \frac{1}{n!}$	*e = 2.7182818284590 45...* – called as *irrational* (as $\boldsymbol{\pi}$), is the mathematical and practical guarantee for never having a linear process identical to another!
while with reference to 1C3:	$p = n!$ and $D_n = !n$ and i all three are *rational*

All equations and definitions above in this section are taken from different sources.

The *permutation* consist all possible process variants, including the real processes as well. The *derangement* consist all variants, apart from the acting *S* sequence, the real process itself. With reference to 2C3, 2C5 and 2C6: there is infinite small, but remaining quantum entropy within each elementary process, which prevents events from *repetition*!

The examples with $\boldsymbol{\pi}$ and $\boldsymbol{e}$ demonstrate the examination of events/processes from all possible angles is obligatory condition in order to find to the correct result.

S. 2.2

2.2
Revolving in relativistic way

Revolving at radius *r* around a stationary centre, might also be understood as being at relative rest while all existing world revolves around us. The experienced conventional view of the circulation means: while the revolving subject keeps its position without rotation, it is at the same time in rotation relative to the centre of the circulation.

Not just the direction of the circulation is the opposite in this relativistic view, but the original centre is the one in rotation relative to the, in this case, stationary subject, as centre of revolution.

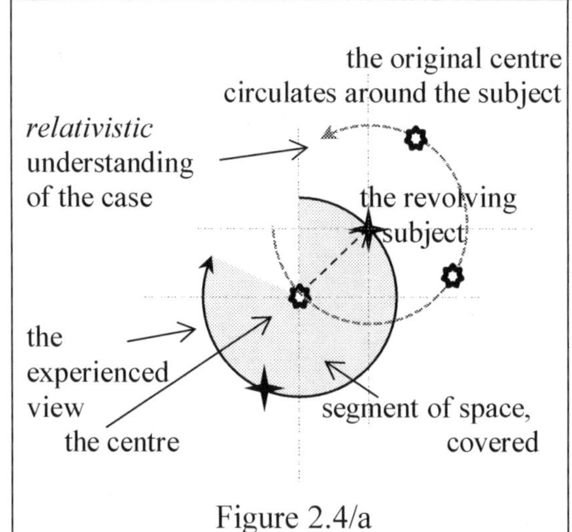

Figure 2.4/a

The closer the two relative centres of the circulation are (the less the radius is), the less is the difference between the information incoming from the quantum space to the two relative centres.

The impacts/information, approaching the subjects in relative circulation is different!

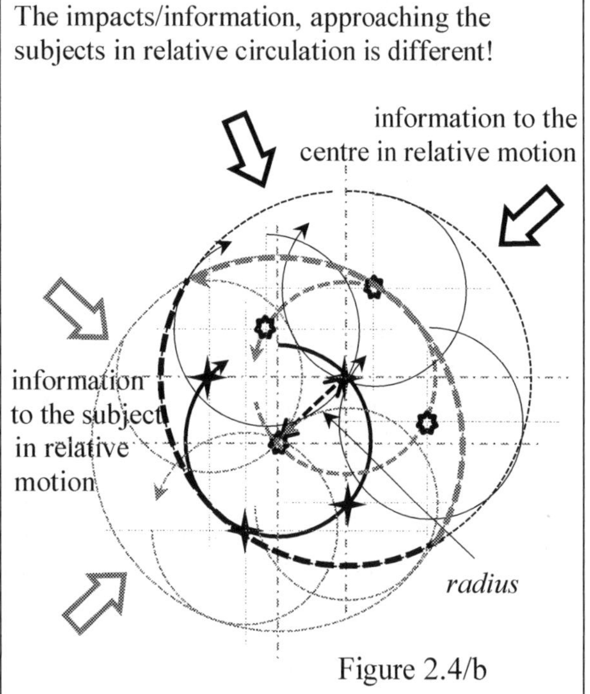

Figure 2.4/b

Fig. 2.4/a

Fig. 2.4/b

The difference between the quantum communications of the two relativistic views of one and the same event is also relativistic: the intensities and also the information coming from the quantum space are different, as the complexity of the relativistic circulation and rotation cover partially different segments of the quantum space. The larger the radius is, the more is the difference, while the proportions remain unchanged.

There are important notes to be taken into the view:

1. The *centre* and also the *subject* – whatever the variant of their relation is – are events.
2. While the events of the relation are separated in space as inputs – in the case of the difference in the intensities and the speed values of the quantum communication – they happen for different length in time and are separated in time as well.
3. Different space-times mean not just different time counts, but also different distances. The radius of the relation is also different, corresponding to the space-time of the measurement.

   The intensity characteristic of the space-times is the *Intensity Quotient*, in fact the quantum drives. For the *Earth* space-time it is $IQ_{Earth} = \frac{c_{Earth}^2}{\varepsilon_g}$; 2D1

   where ε_g is the intensity coefficient of gravitation.
4. $dt_x = \frac{dt_o}{\sqrt{1-\frac{v^2}{c_x^2}}}$; The events of the relative examination have technically equal v values, but different time counts and space positions. As value c_x varies, the information differs in space and time anyway. 2D2

Fig. 2.5

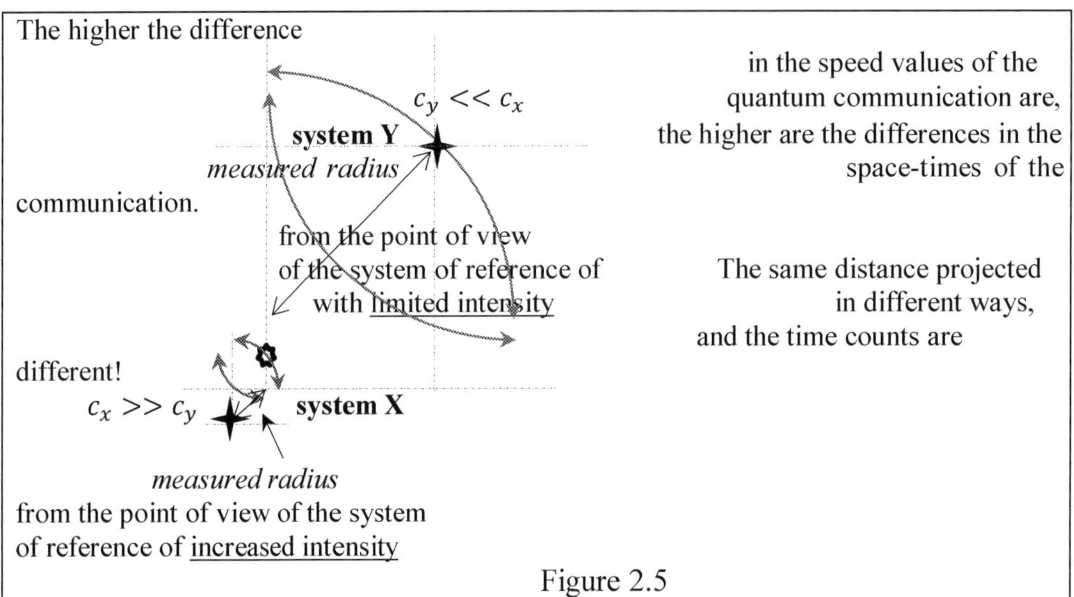

Figure 2.5

In the *system of reference X* the intensity is high, the time count and the measured distance are short; the time is passing slowly; while in the *system of reference Y* the time count is speeded up, the space coordinates are measured at lengthy distances; the information are approaching the space coordinates from far away.

The physical values of the speed are equal, but their measurements and expressions are different, function of the space-time.

5. Events in relative relations appear in different ways.

   The examination of the universe from the surface of the *Earth* means that all those planets, subjects and configurations on the sky have their own space-time, very different from each other, not just in space, but also in time, and different than our space-time on the surface of the *Earth.*

S. 2.3

2.3
The continuity of the elementary processes and the case of the transmutation

Diag. 2.2

The decreasing intensity of the quantum impacts from the inflexion up to the fully extended status = = *proton* and *electron* processes	The increasing intensity of the quantum impact form the fully extended electron process to inflexion = = *neutron* process
sphere symmetrical expanding acceleration	sphere symmetrical accelerating collapse
electron process proton process inflexion Diagram 2.2/a	fully extended electron process status driven by the electron process neutron process inflexion Diagram 2.2/b

from $\frac{dmc_x^2}{dt_n}\sqrt{1-\frac{(c_x-i_c)^2}{c_x^2}}$; to $\frac{dmc_x^2}{dt_n}\left[1-\frac{(c_x-i_x)^2}{c_x^2}\right]$	The two ends of the direct elementary process are connected by the anti-direction of the elementary process.	2E1 2E2

The anti-processes are mandatory components of the elementary evolution. They are the keys of the fluency of the elementary processes.
This means permanent impact of the components of the elementary process. It can be interpreted the following way:

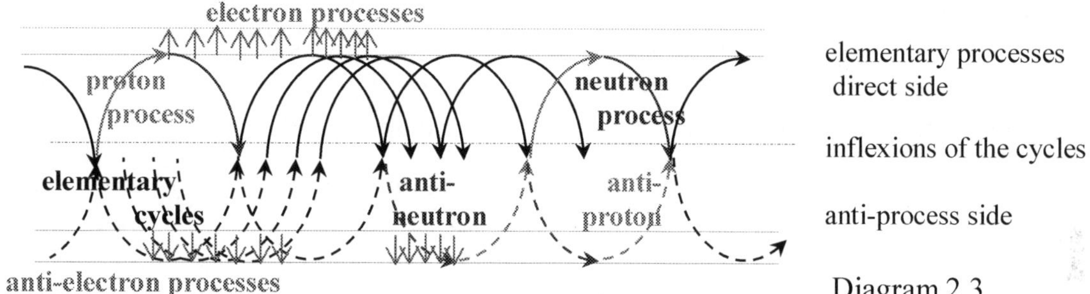

Diagram 2.3

Diag. 2.3

The quantum drives of the anti-electron processes are equal for all elementary processes, with the variety of the speed of the quantum communication and the intensity coefficient of the electron process. Therefore the final intensity of the anti-proton process and the intensity capacity of the proton process (and this way the intensity of the electron process) depend on the intensity impact of the anti-electron process drive.

The higher the quantum speed of the communication of the elementary process is, the more is the generating anti-electron process surplus, the established quantum membrane and the higher is the number of the elementary cycles.

With reference to Diagram 1.2

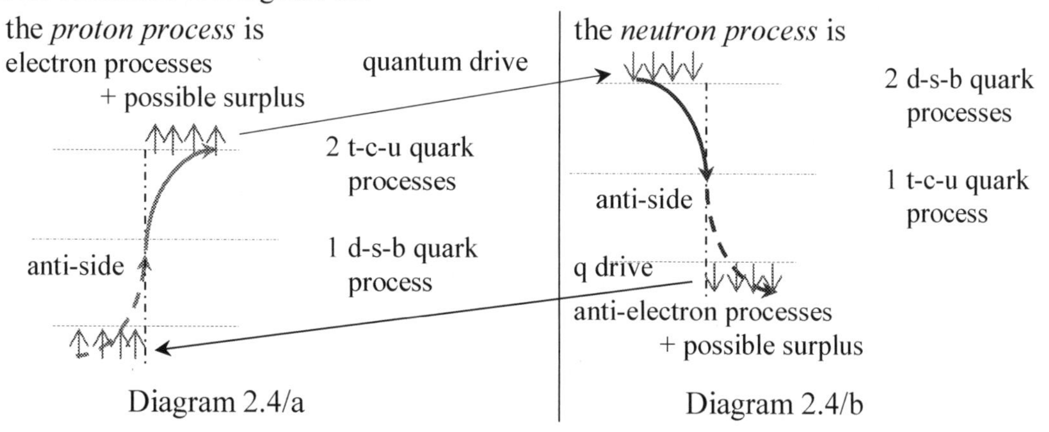

Diagram 2.4/a Diagram 2.4/b

Diag. 2.4

Through the intensity exchange on the direct and also on the anti-process sides the elementary processes are in full balance.

t=172.44 GeV; c=1.27 GeV; u=2.4 MeV; d=4.8 MeV; s=95 MeV; b=4.18 GeV 2E3

The intensities of the anti-electron processes of all elementary processes of the elementary evolution are equal to the intensity of the quantum impact of the *gravity*. The surplus of the intensities of the anti-electron processes, generated in the elementary processes might be directed in certain way and focused to a certain subject.

The intensity of the expansion of the anti-neutron process generates the anti-electron process quantum drive, in numbers of

2E4 $$(n_s + n_e)\frac{c_x^2}{\varepsilon_{x-}};$$ n_s - the number of the surplus; n_e - the number of the quantum drive of the anti-proton process.

The higher the periodic number of the elementary process is, the higher are ε_{x-} the intensity coefficient of the anti-electron process, the generating surplus and the quantum speed of the elementary communication.

The increased quantum speed of the communication means an increased space for the equal time count, or the same space coordinate for the reduced time count.

The speeding up of the elementary process results in the intensity increase of the electron process drive, the neutron collapse and the increase of the surplus of the anti-electron processes.

2E5 $$e_e = \frac{dmc_x^2}{dt_i\varepsilon_x\sqrt{1-\frac{v^2}{c_x^2}}}\left(1-\sqrt{1-\frac{(c_x-i_x)^2}{c_x^2}}\right); \quad e_{e-} = \frac{dmc_x^2}{dt_i\varepsilon_{x-}\sqrt{1-\frac{v^2}{c_x^2}}}\left(1-\sqrt{1-\frac{(c_x-i_x)^2}{c_x^2}}\right);$$

But the direct side is about $c_x^2 \cdot \varepsilon_x = \text{const}$; and the anti-side it about $\frac{c_x^2}{\varepsilon_{x-}} = \text{const}$.

Therefore it would be impossible for keeping the elementary process without change,

2E6 $$\text{as } c_x^2 \cdot \varepsilon_x \neq c_x^2 \cdot \varepsilon_x \cdot \sqrt{1-\frac{v^2}{c_x^2}} \text{ and } \frac{c_x^2}{\varepsilon_{x-}} \neq \frac{c_x^2}{\varepsilon_{x-}\sqrt{1-\frac{v^2}{c_x^2}}}.$$

At the same time the speeding up is a real process.

It can only be managed at the elementary level only if the number of the cycles remains the same, but with the increase of the intensity. The equation in 2E5 in this case is written in the following format:

2E7 $$e_e = n_x\frac{dmc_x^2}{dt_i\varepsilon_x}\left(1-\sqrt{1-\frac{(c_x-i_x)^2}{c_x^2}}\right); \quad e_{e-} = n_x\frac{dmc_x^2}{dt_i\varepsilon_{x-}}\left(1-\sqrt{1-\frac{(c_x-i_x)^2}{c_x^2}}\right);$$

The number of the cycles happens for shorter time count and the surplus of the anti-electron processes becomes increased. The acceleration is increasing the intensity of the elementary evolution. There is a point however when the acceleration cannot be managed by the increase of the cycles any more. The speeded up elementary cycles result in the conflict themselves. The conflict ends up with the generation of a new elementary process in line with the equations in 2E5: resulting in an elementary process (*transmutation* in fact) with increased quantum speed and electron process intensity (with less numerical value of the coefficient ε_x). A step backwards in the line of the elementary evolution.

2.4 The meaning of the *Euler's* identity the inflexion and the indeterminacy

S. 2.4

$\boldsymbol{e^{i\pi} + 1 = 0;}$ or 2F1

$\boldsymbol{e^{i\pi} = -1;}$ or 2F2

$\boldsymbol{-e^{i\pi} = 1;}$ 2F3

i – is an imaginary number defined as the number we get by taking the square root of (-1).

This is the equation, the one the mathematician Benjamin Pierce told about:
"It is surely true, it is absolutely paradoxical; we cannot understand it, and we do not know what it means. But we have proved it, and therefore we know it must be the truth."

Let us see what it may mean if using the process based approach.

In the terms of physics it shall mean the unity of the processes and the anti-processes!
There is no process without its anti-process.
The numbers on the both sides of the equation are in fact equal to, and mean

$e^{i\pi} = \cos\pi + i\sin\pi =$

$= 1 + i\frac{\pi^1}{1!} - \frac{\pi^2}{2!} - i\frac{\pi^3}{3!} + \frac{\pi^4}{4!} + i\frac{\pi^5}{5!} - \frac{\pi^6}{6!} - i\frac{\pi^7}{7!} + \frac{\pi^8}{8!} + \cdots;$

the processes never become completed!

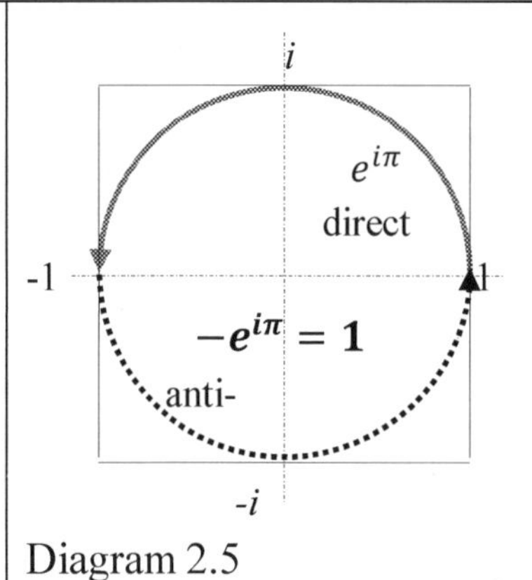

Diagram 2.5

Diag. 2.5

$\boldsymbol{\pi}$ – the message is: there is no way for repeating any cycle of events!

$\boldsymbol{e}$ – the unique number such that the value of $\boldsymbol{e^x}$ is equal to the slope of its tangent line for every *x*.
the message is: going right ahead, in direct line, without any cycle, the reasons and the goals of the events are always different!

$\boldsymbol{i = \sqrt{-1}}$ the message is: for finding the truth it needs always a double-sided approach!
The root cause of the events comes out always from this imaginary relation of:

$$(1) * (-1) = -1 \qquad \text{2F4}$$

The first two numbers are called as *irrational*, the third as *imaginary* numbers.
The message ***Euler's equation*** (the three together) is: the inevitability of the *relativistic* views. The existence of the event and its reverse in parallel!

At the same time the *number line* with the so called *real numbers* as such *does not exist,* or only exists in a specific format.
There is no way to find and define any number in the number line, an integer, or in any fractions. There might always be named other numbers, closer, approaching, or leaving the one to be defined.

The example is:
Number 3, as such cannot be defined.
There might always be other numbers to be found from both directions, either from below, as 2.9999999…9 continuing the approach for infinity up to the supposed 3 or from above, as 3.0000000000…1 also continuing it also for infinity. The closeness might be of infinite low value, but never making possible to reach the exact number 3.
And there is no difference, is it an integer like: 0; 1; 123; 4567, or any fraction as 2.345 or 1234.56789 and so on.

Each number – integer or in fractions, no difference – may fulfil the function of the *zero*, the officially appointed position of the inflexion, the change of the directions.

The numbers on the number line exist only as *inflexions*: the only events in physics without time, without duration! Exempt of course of those – the $\boldsymbol{\pi}$, the $\boldsymbol{e}$ and the $\boldsymbol{i}$ – called in strange way *irrational***,** in spite they are the only ones about the reality of the events: the *indeterminacy*!

Indeterminacy does not mean the miss of the reasons and the lack of the consequences at all! In the contrary! It means events happen in time with directions, initiations, functions, purposes and motivations, without repetitions, but most importantly in time and space.

This is the proof in mathematics:

- The rest, as the status of the matter does not exist in physics.
- Everything in the *Universe* is in motion, event in *time*.
- The so called real numbers in mathematics are just the tools for the definition of the current position of the motion and the event in the *space*.
- It proofs the unity of the *time* and the *space*.

3
Magnetic field and relativity

1. If a wire (or any other subject) with elementary processes inside in motion is, the intensities of the elementary processes inside become increased.
2. The collapse of the neutron processes of the wire (or any other subjects) in this case is driven by the electron processes of increased intensity.
3. The anti-electron processes are controlling the identities of the elementary processes and keep it unchanged. The consequence of the intensity increase of the neutron collapse is the intensity increase of the anti-electron processes, resulting in the generation of additional an intensity surplus.
4. The intensities of the anti-electron processes are equal for all elementary processes. The intensity increase means the increase of the number of the quantum impacts of the anti-electron processes. There is a surplus in the majority of the elementary processes, therefore the motion in this case means the generation of an additional surplus.
5. The intensities of the elementary processes might obviously change in the other direction, becoming less value as well. The less intensity in this case is the result of the decrease of the intensity of the electron processes. The most usual way for the correction is the reduction of the release of the standard surplus of the anti-electron processes. The other, more radical way is looking for external elementary assistance that is initiating external elementary communication/s. There is no way for increasing the intensity on its own.
6. The no need for correction would mean any change in the intensity and the generating surplus of the anti-electron processes.
7. The generating additional intensity surplus of the anti-electron processes, the result of the motion, with reference to points 2 and 3, in this case is released.

Ref. p.2 p.3

8. The release of the additional surplus brings down the intensities of the electron processes and the elementary process returns to its standard status.
9. The release of the surplus of the anti-electron processes means a magnetic impact, which is establishing the electromagnetic field of the elementary process.
10. The release of the additional surplus increases this magnetic impact.
11. If there is a wire within the electromagnetic field of another elementary process, the two electromagnetic fields communicate. As the consequence of the motion, the communication in this case is strengthening, the result of
    - the impact of the released normal and additional surplus; and
    - the impact of the normal surplus of the electromagnetic field of the other elementary process.

Ref. p.2

12. There is a conflict, developing between the two electromagnetic fields.
13. While the conflicting two magnetic fields usually establish a balance, each new effect in the relation, like the sectioning of the magnetic fields by the other fields, result of the motion of the wire destroys the balance and renews the conflict.
14. The conflict of the acting electromagnetic fields increases the intensities of the quantum impacts of the anti-electron processes in both elementary processes.
    15. The increase of the intensities of the anti-electron processes results in the increase of the intensities of the quantum impacts of the electron processes within the wire.
16. This increase is equivalent to the increase of the number of the acting electron processes (above its standard value).
17. The increase of the intensity of the electron processes either increases the internal temperature or the generating additional electron process impacts propagate away.
18. The conventional term of the propagating *blue shift* quantum impacts of the electron processes towards other elementary processes with less intensities, is the *current, electricity.*
    19. If the conflict of the electromagnetic fields is permanent, the flow of the electron process *blue shift* quantum impacts, the *current* is also permanent.
20. The *electricity* is not about the motion of the electrons, rather the propagation of the quantum impacts of the electron processes; the hand-over of the *blue shift* impacts by the elementary processes.
    21. The term of the intensity difference between the elementary processes of the generation and the receipt of the *blue shift* quantum impacts is: *voltage.*
    22. High intensity potential generates a flow of *blue shift* quantum impact of massive *density,* with high value *electric current.*
23. The electric *current,* circulating within the coils of a solenoid around an iron core generates electromagnetic field within the core.
24. The electromagnetic field itself is also the result of a kind of relativistic impact:
    25. The quantum impacts of the electron processes propagate within the wire of the *solenoid.* This is in fact acceleration in the direction of the radius of the solenoid. *The acceleration is increasing the intensity of the electron processes within the wire.*

Fig. 3.1

Figure 3.1

26. The difference between points 24 and 3 is that the increase of the intensity here is the result of the circulation/acceleration of the electron process *blue shift* impact (the current) itself, while in point 3 the increase is coming from the motion of the wire.

Ref. p.3

27. The increasing this way intensity of the electron process generates additional *blue shift* surplus within the wire of the coils.
28. The intensity increase of the anti-electron processes within the coils is impacting the *environment* and the elementary processes of the *core of the solenoid* as well.

29. The unit values of the quantum impacts (the intensities) of the anti-electron processes are equal with the quantum impact of the gravity.
30. The impact of the anti-electron process surplus towards both, the *environment* and the *core* of the solenoid generates conflict.
    31. The quantum impacts of the anti-electron processes, the electromagnetic field itself have a certain direction within the core of the solenoid. (Discussed in details in Section 3.4.) Ref. S.3.4
    32. The value of the intensity of the electromagnetic field of the core depends on the intensity values of the quantum impacts of the anti-electron processes, generated by the solenoid. And the intensity of the anti-electron processes depends on the value of the current within the coils of the solenoid.
33. In order to keep the standards of the elementary processes of the core of the solenoid, the surplus of the generating quantum impacts of the anti-electron processes shall be released at both ends (discussed in details later in Section 3.1.) Rel. S.3.1
    34. The impact of the generating anti-electron process surplus is acting through the casing of the solenoid as well. As the surface of the casing is larger, the density of the impact is of less value.
35. The elementary processes have their own electromagnetic impact. The intensities of the impacts however are different. The *Iron*, the *Iron-oxides*, the *Carbon steel*, the *Cobalt, Magnesium* and the *Nickel* elementary processes have the most intensive electromagnetic impacts. The reason of the differences in the intensities is the difference in the density of the elementary processes, discussed in Chapter 5. Ref. Ch.5
    36. The effects of the communication of the elementary processes with magnets have three formats;
        - attraction: *ferromagnetic*, like Iron (with *strong*); *paramagnetic* like Magnesium (with *very weak*) impacts; and
        - repel: *diamagnetic*, like Silicon (with very weak).
    37. The relation of the elementary processes depends on the intensity of the electron processes. Table 3.1 below is about the relation of elementary processes to the magnetic impacts.

1	H	Dia	31	Ga	Dia	61	Pm	N/A	91	Pa	N/A
2	He	Dia	32	Ge	Dia	62	Sm	Para	92	U	N/A
3	Li	Para	33	As	Dia	63	Eu	Para	93	Np	N/A
4	Be	Dia	34	Se	Dia	64	Gd	Ferro	94	Pu	N/A
5	B	Dia	35	Br	Dia	65	Tb	Para	95	Am	N/A
6	C	Dia	36	Kr	Dia	66	Dy	Para	96	Cm	N/A
7	N	Dia	37	Rb	N/A	67	Ho	Para	97	Bk	N/A
8	O	Para	38	Sr	Para	68	Er	Para	98	Cf	N/A
9	F	N/A	39	Y	N/A	69	Tm	Para	99	Es	N/A
10	Ne	Dia	40	Zr	Para	70	Yb	N/A	100	Fm	N/A

11	**Na**	Para	41	**Nb**	N/A	71	**Lu**	N/A	101	**Md**	N/A
12	**Mg**	Para	42	**Mo**	Para	72	**Hf**	N/A	102	**No**	N/A
13	**Al**	Para	43	**Tc**	N/A	73	Ta	N/A	103	**Lr**	N/A
14	**Si**	Dia	44	**Ru**	Para	74	W	Para	104	**Rf**	N/A
15	**P**	Dia	45	**Rh**	Para	75	Re	N/A	105	**Db**	N/A
16	**S**	Dia	46	**Pd**	Para	76	Os	Para	106	**Sg**	N/A
17	**Cl**	Dia	47	**Ag**	Dia	77	Ir	Para	107	**Bh**	N/A
18	**Ar**	Dia	48	**Cd**	Dia	78	Pl	Para	108	**Hs**	N/A
19	**K**	N/A	49	**In**	Dia	79	Au	Dia	109	**Mt**	N/A
20	**Ca**	Para	50	**Sn**	Para	80	Hg	Dia	110	**Ds**	N/A
21	**Sc**	N/A	51	**Sb**	Dia	81	**Tl**	Dia	111	**Rg**	N/A
22	**Ti**	Para	52	**Te**	Dia	82	**Pb**	Dia	112	**Cn**	N/A
23	**V**	N/A	53	**I**	N/A	83	**Bi**	Dia	113	**Uut**	N/A
24	**Cr**	ferro	54	**Xe**	Dia	84	**Po**	N/A	114	**Uuq**	N/A
25	**Mn**	Para	55	**Cs**	N/A	85	**At**	N/A	115	**Uup**	N/A
26	**Fe**	**Ferro**	56	**Ba**	Para	86	**Rn**	N/A	116	**Uuh**	N/A
27	**Co**	**Ferro**	57	**La**	N/A	87	**Fr**	N/A	117	**Uus**	N/A
28	**Ni**	**Ferro**	58	**Co**	Para	88	**Ra**	N/A	118	**Uuc**	N/A
29	**Cu**	Dia	59	**Pr**	N/A	89	**Ac**	N/A			
30	**Zn**	Dia	60	**Nd**	Para	90	**Th**	N/A			

Table 3.1

Table 3.1

S. 3.1

3.1
Electromagnetic impact

Ref. Table 3.1

38. The majority of the elementary process generates measurable electromagnetic field. The electromagnetic field is impacting the anti-electron processes of other elementary processes within a certain distance.
39. With reference to the Table 3.1, the elementary processes with *ferromagnetic* characteristics are the ones with more dynamic reaction on the impact.
40. The electromagnets with solenoids generate magnetic fields and have two open ends, for keeping their internal balance stable under the impact of the solenoid.
41. The subject on the Figure 3.2 is under the impact of the electromagnetic field of the core. The *South* end is one of the releases.
42. The electromagnetic impact of the *South* end generates conflict and increases the intensity of the electron processes within the elementary processes of the subject.
43. This results in the additional increase of the quantum impact of its anti-electron processes. The developing additional surplus shall be released in order to keep the identity of the elementary process unchanged.
44. The quantum impacts of the developing surplus of the anti-electron processes of the subject under the magnetic impact are released in all directions.

45. There is a conflict, developing between the quantum impact of the released anti-electron processes of the *South* end and the elementary processes. One part of the conflict disappears in the open space; the other part of the conflict weakens and/or eats off the anti-electron process quantum impacts of the subject.

46. The effect of the electromagnetic impact on the paramagnetic and ferromagnetic subjects is attraction; South end:

47. The electromagnetic field of the core is the one initiating the magnetic impact.
48. The electromagnetic impact increases the intensity of the quantum membrane of the anti-electron process of the subject.
49. The higher the density of the electromagnetic field is, the higher is its magnetic impact.

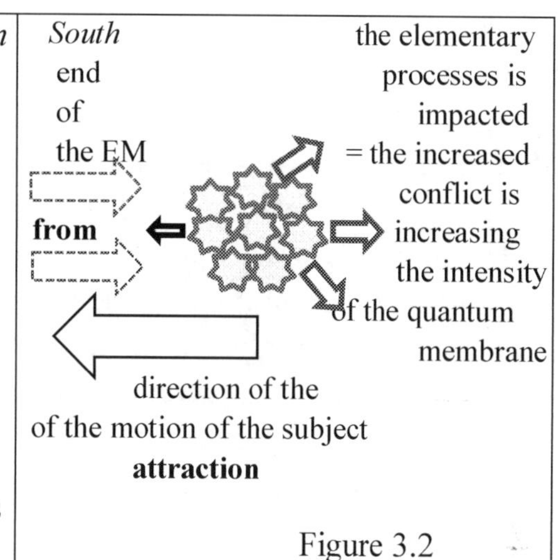

Figure 3.2

Fig. 3.2

50. The developing conflict results in the misbalance of the elementary process of the subject. The subject is looking for the way for re-establishing the balance and the active quantum impacts of the electromagnet moves the subject towards the *South*, with reference to point 4 – in order to restore the balance.

Ref. p.4

51. The balance includes both components of the relation, the magnet and the subject as well. The magnetic impact of the electromagnet is released now through the elementary processes of the subject.
52. The impact of the *North* end is similar, but the explanation is different and more obvious.

53. The effect of the electromagnetic impact on paramagnetic and ferromagnetic subjects is also attraction; North end.

54. There is a deficit in the impacts of the anti-electron processes at the *North* end.
55. The deficit becomes covered by the released quantum impact of the subject. It is weakening at this side.
56. The overwhelming impact of the released of the other direction moves the subject towards the *North* end. It is attracted.

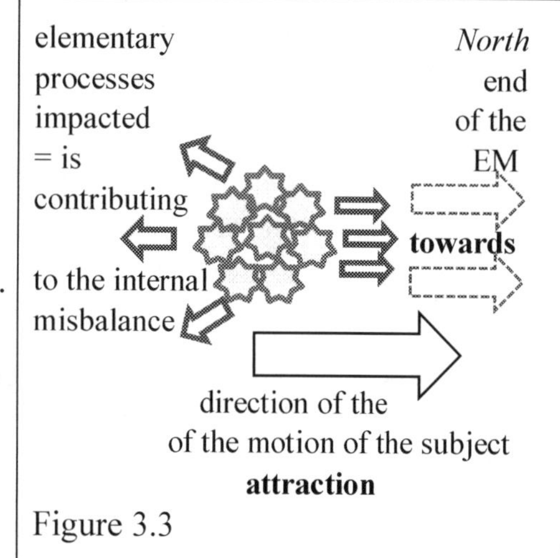

Figure 3.3

Fig. 3.3

57. The electromagnetic effects of the two ends of the electromagnets (and also the natural magnets) are different, but the relations and the impacts towards the external subjects are similar. Both ends are attracting.

58. The attractions to both, towards the *South* and *North* ends improve and restore the internal balance of the subjects.
59. The relations between two electromagnets to each other and between two natural magnets to each other are different: the similar ends repel, the different ends attract each other.
60. Electromagnets and the natural magnets are *repelling* each other at their similar ends, and *attracting* each other at their different ends – for establishing their balance.
61. The difference between the *paramagnetic* and *ferromagnetic* subjects is in the *densities* of the elementary processes.
62. In the case of *diamagnetic* elementary processes, there is no conflict between the *South* and the *North* ends of the electromagnet and the subject; the magnetic impact weakens the quantum membrane of the elementary process. The subjects receive *dynamic* (once a time acting) *repelling impacts*, rather than permanent ones.
63. The positive and the negative ends of the electromagnets depend on the wiring of the solenoid, to be discussed later in Section 3.3.

S. 3.2

3.2
Generation of electric current, alternate current

64. If there is no way for the subject to move towards the electromagnet, the developing surplus of the electron process *blue shift* quantum impact – generates conflict.
65. If a wire, in motion, driven by external force within an electromagnetic field is, there is no chance reaching a balanced status. It results in permanent flow of the electron process *blue shift* impact, *current,* within the wire.
66. In the case the wire is rotating and sectioning this way the electromagnetic field time to time in different directions, the direction of the generating current is also changing.
67. The direction of the generating *current* within the wire depends on the direction of the motion, the direction of the sectioning of the electromagnetic field.

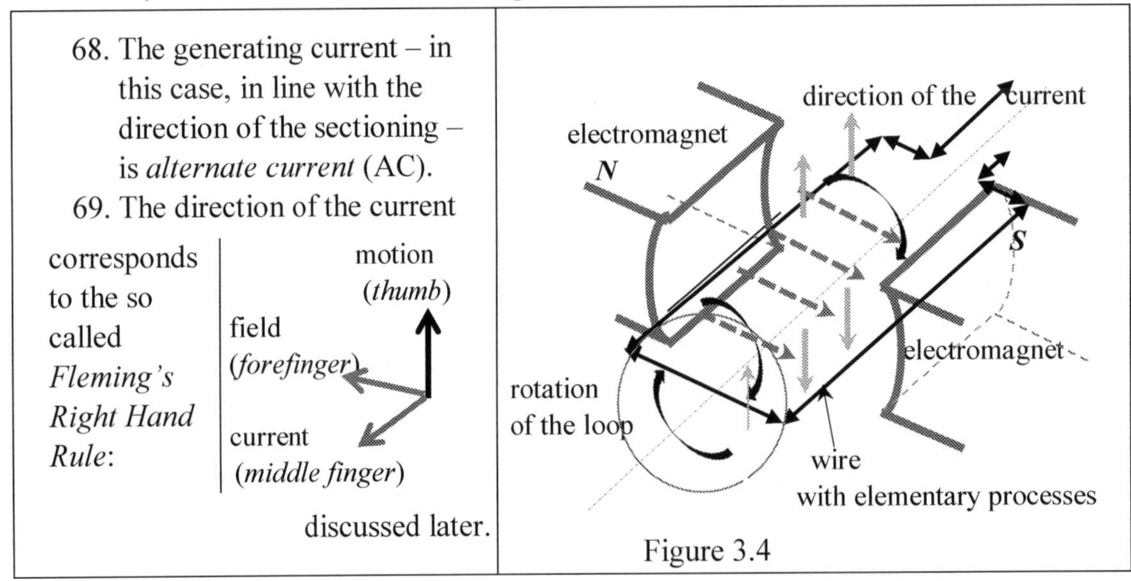

Fig. 3.4

Figure 3.4

3.3 The directions of the magnetic field within the earth and above the surface

S. 3.3

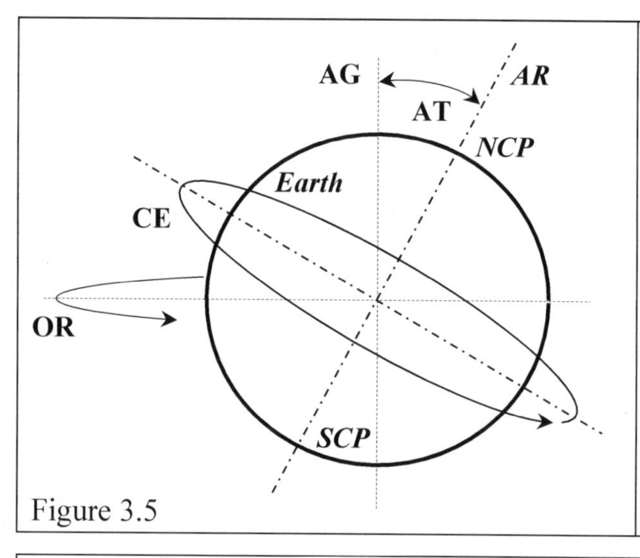

Figure 3.5

The axis, connecting the ***Northern Celestial Pole*** (*NCP*) with the ***South Celestial Pole*** (*SCP*) is the ***Axis of the Rotation*** (AR). The ***Celestial Equator*** (*CE*) is perpendicular to the *AR*. There is *Axis Tilt* (*AT*) between the *AR* and the ***Axis of the Geometry*** (*AG*), which is perpendicular to the ***orbit of revolution of the Earth*** (*OR*) around the *Sun*. The impact of the *Sun* is constant.

Fig. 3.5

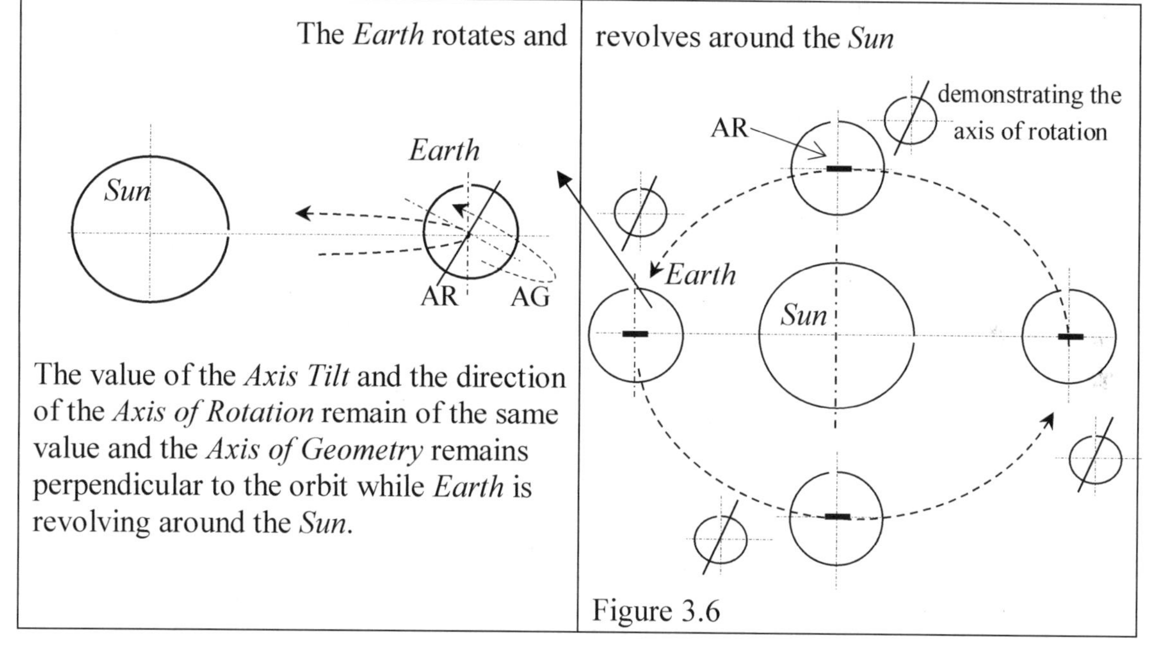

Figure 3.6

Fig. 3.6

3.3.1. The reason and the proofs of the generation of the magnetic field

S. 3.3.1

a) The electromagnetic quantum impacts of the *Sun* increases the intensities of the elementary processes of the *Earth*. The increase of the intensities is the result of the increasing conflicts of the electron processes. The increased intensity means the generation of the additional surplus (the surplus) of the anti-electron processes within the *Earth*.

b) The quantum impact of the *Sun* is *external*, quasi constant and permanent. It cannot be influenced. The elementary processes of the *Earth* are the ones, which shall react and give permanent response in order to guarantee the standards of the elementary processes of the *Earth* and the quantum impact of *gravity* of the *Earth*.

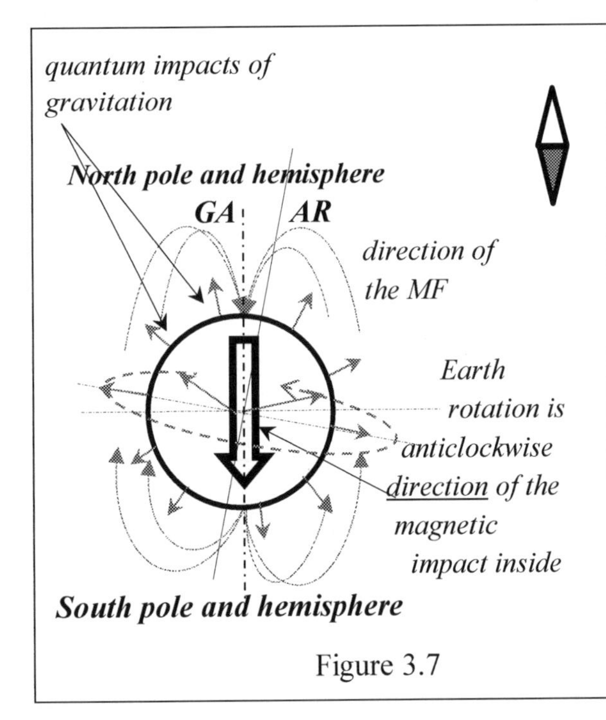

Figure 3.7

Fig. 3.7

The reason of the generation of the difference in the magnetic impacts of the *Earth* is the *Axis Tilt* between the *Axis of Rotation* and the *Axis* of *Geometry*.

Because of the different tangential speed values, the elementary processes of the *Earth* become of different intensities; because of the rotation they are getting closer to and farer from the quantum impact of the *Sun*.

The local points have time to time increased or decreased intensities.

The magnetic lines above the *Earth* surface are the continuation of the inside direction and are closed at the geographic *North* pole establishing the magnetic balance of the *Earth*.

c) The intensities of the quantum impacts of the anti-electron processes of the elementary processes of the *Earth* are of equal values, but the speed values of their quantum communication vary.

d) For the compensation of the external load of the *Sun*, the acting number (the density) of the quantum impacts of the anti-electron processes of the *Earth* shall also be of increased values within the impacted areas. The increase of the density causes conflicts, resulting in the increase of the temperature above the surface.

e) The rotation of the *Earth* is one of the consequences of the conflict. The circulation of the *Earth* around the *Sun* is the consequence of the rotation. *The direction of the rotation coincides with the direction of the circulation.* The *Earth* simply goes ahead as the circulation dictates.

f) There is a certain asymmetry anyway in the geometry of the *Earth*, since the increasing tangential speed values in the direction towards the *Celestial Equator* are increasing the intensities of the elementary processes.

g) The increased tangential speed, contrary to the quantum impact of the *Sun*, is an internal impact. The consequence is a slight deformation of the geometry of the sphere symmetrical expanding acceleration. The speeded up elementary processes result in increased intensity of the expanding acceleration. And the geometry is responding to the impact.

h) The quantum impact of the *Sun* has its permanent direction, corresponding to the orbit of the *Earth*, perpendicular to the *Axis of Geometry* of the *Earth*. The load, relative to the *Axis of Rotation* is not symmetrical. While the positions of the symmetrical segments and points at the surface relative to the geometry are similar, their external quantum impacts are different relative to the *Axis of Rotation*.

i) The dominancy of the impact, as consequence of the rotation is directed toward the *southern* hemisphere, below the *Celestial Equator.* Therefore the generating – by the elementary processes inside the *Earth* – anti-electron process surplus is driven towards the *southern* hemisphere for the compensation of the increased loss, result of the increased conflict above the surface on the *southern* hemisphere .
j) The drive is the difference in the intensity values. The generating conflict of the increased intensity all above the surface of the *southern* hemisphere needs compensation from the *northern* hemisphere and the surplus is given off towards the *Southern Pole.*
k) The *magnetic* impact is no other than the internal flow of the quantum impacts of the anti-electron processes from the *northern* hemisphere to the *southern* for balancing the increased loss of the increased conflict towards the *southern* pole!
l) The given off anti-electron process surplus through the *southern hemisphere* and the *southern pole* is missing from the elementary balance of the *Earth.* Therefore the surplus of the quantum impact of the anti-electron processes of the *southern* hemisphere returns back through the *northern hemisphere* and the *northern pole* above the *Earth* surface.
m) The *magnetic* impact means the *blue shift* impact of the anti-electron processes below and above the *Earth* surface.

The proofs of the dominant quantum impact of the *Sun* on the *southern hemisphere* are given in the following sections.

3.3.2. Relativistic view is the key of the assessment

S. 3.3.2

If the *Earth* rotates and revolves around the *Sun,* it might also be understood in its relativistic way that the *Sun* has its specific motion relative to the *Earth.*

The rotation of the *Earth* around its *AR* is anticlockwise. And the circulation of the *Earth* is also anticlockwise.
This means, the *Sun* is revolving around the *Earth* in relativistic terms, in clockwise direction.

The relation of the motion of the *Sun* and the *Earth* on Figure 3.6, may also be presented, as Figure 3.8 demonstrates it.

The position of the *Earth* is fixed, but rotating around the *AR,* while the *Sun* is revolving around. The direction of the *AR* and the tilt of the rotation remain unchanged.

This assumption does not change at all the relation.

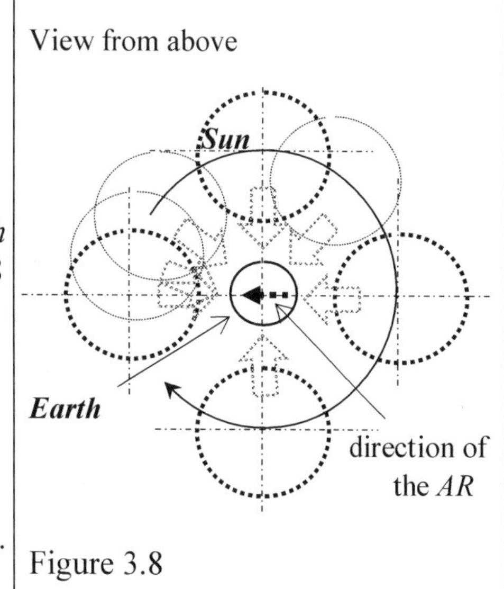

Figure 3.8

Fig. 3.8

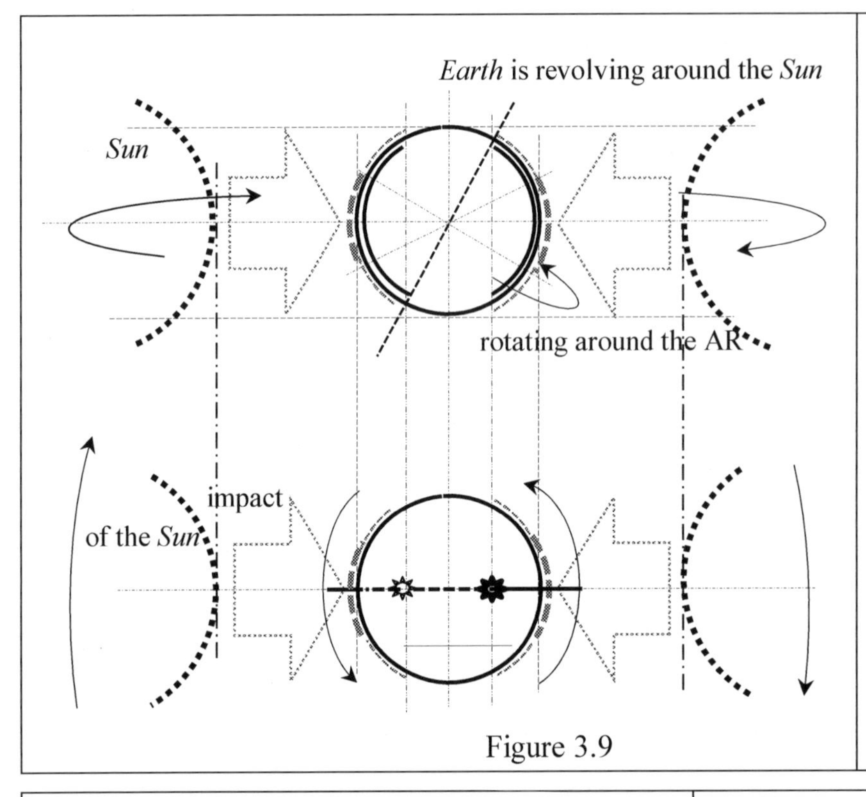

Figure 3.9

The supposed relativistic concept that the "*Earth* is rotating while the *Sun* is revolving around" does not give the definition of the magnetic field within the *Earth* in the attached form.

It seems the load from the *Sun* is ***symmetrical***.

Therefore each variant is also valid in its reversed understanding as well.

Fig. 3.9

If it is taken that the *Earth* is in position fully fixed, even without rotation, the relativistic motion of the *Sun* around, because of *Axis Tilt*, is a spiral.

The discrete positions of the *Earth* and the *Sun* on Figure 3.11 demonstrate the motion of the *Sun*. Each position is representing the rotation of the *Earth* as well = the relative motion of the *Sun* is a spiral.

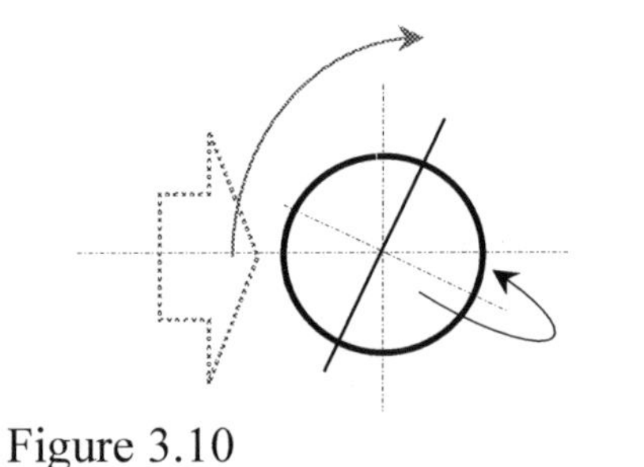

Figure 3.10

Fig. 3.10

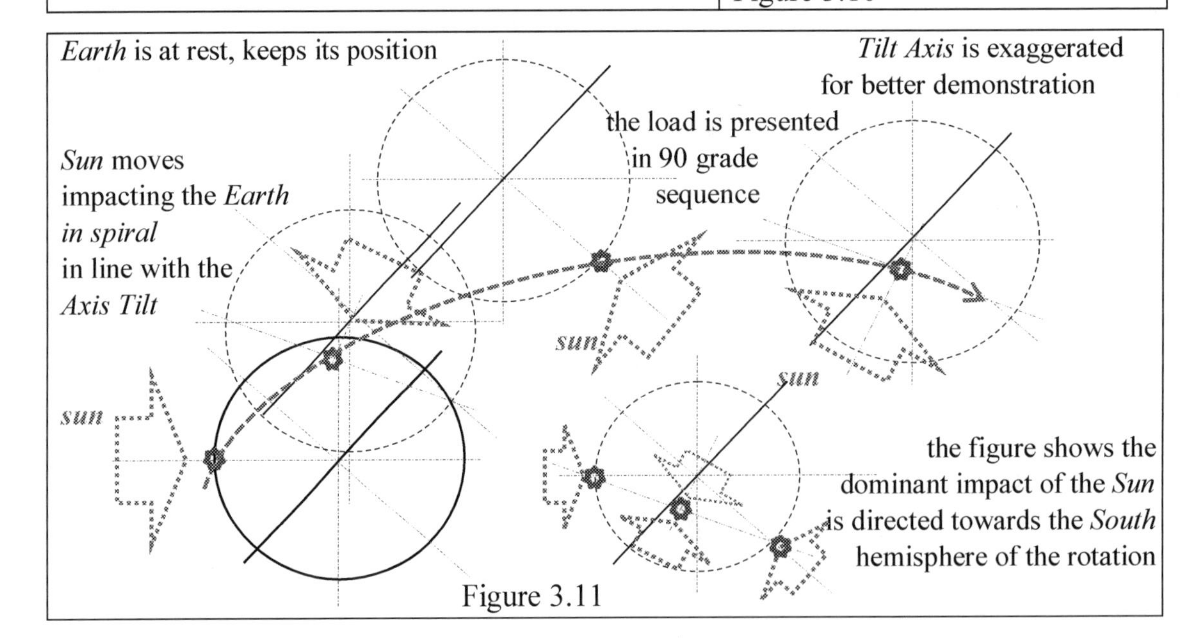

Figure 3.11

Fig. 3.11

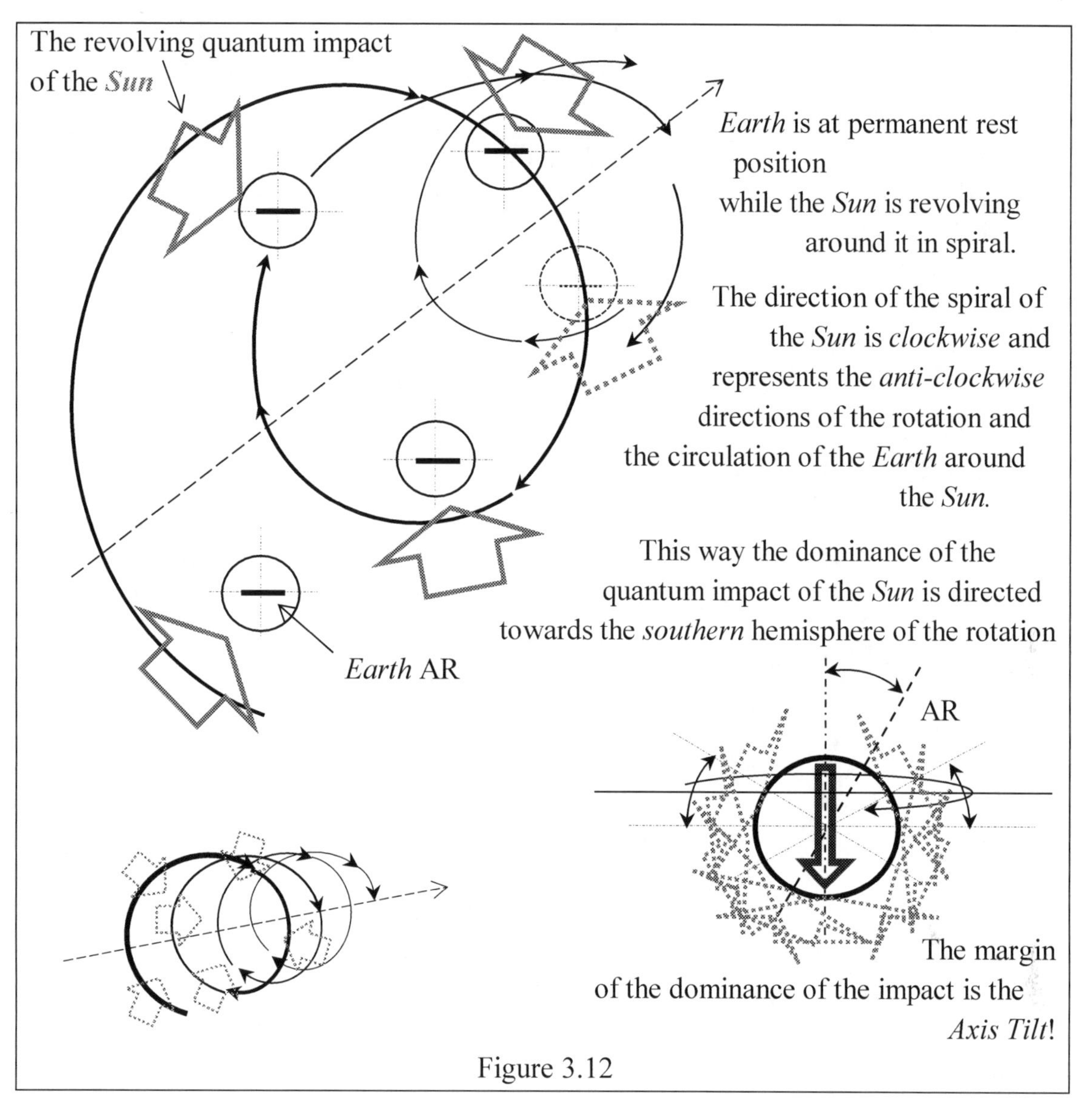

Figure 3.12

Fig. 3.12

The global relativistic view is the one only, which brings us closer to the correct answer. If we take that the position of the *Earth* is motionless – with no rotation and no circulation – all impacts are the consequence of the motion of the *Sun*.

As the electromagnetic quantum impacts of *the Sun* towards the *southern hemisphere of the rotation* is the dominant ones, the *southern hemisphere* is the one where the resulting conflict increases the intensities and the densities of the quantum impacts of the balancing anti-electron processes of the elementary processes. The loss generated by the conflict marks out the direction of the propagation of the anti-electron process quantum impacts from the geographic *North* to the geographic *South* to within the *Earth* for re-establishing the balance.

As final conclusion it can be stated, that the magnetic field inside the *Earth* is directed towards the *South Pole.* There are three points establishing the direction:

- the existing difference of the positions of the axis of the rotation relative to the axis of the geometry of the *Earth*, the tilt of the rotation; and as consequence

- the dominance of the quantum impact of the *Sun* directed towards the *southern* hemisphere; it drives, for the compensation, the anti-electron process quantum impacts of the elementary process within the *Earth* towards the *South Pole;*

Ref. Fig. 3.7

- for establishing/restoring the internal magnetic balance of the *Earth* the quantum impact is directed back to the geographic *North Pole* above the surface, as it is presented on Figure 3.7.

S. 3.4

3.4
The Right Hand Grip Rule (RHGR) and the Right-Hand-Rule (RHR)
Reasons, concerns and consequences

S. 3.4.1

3.4.1. Right Hand Grip Rule

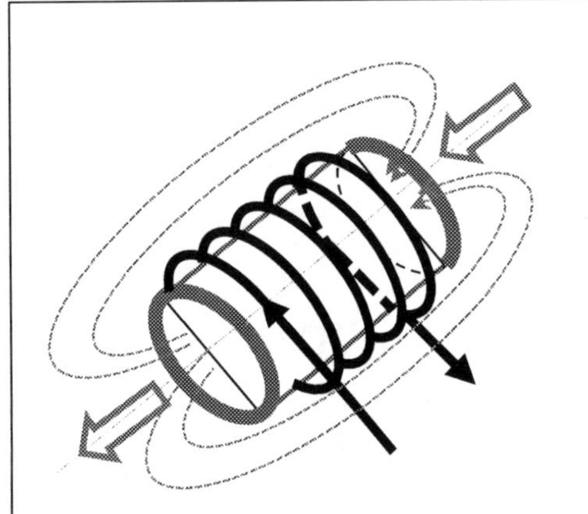

The *current* is flowing in the direction of the arrows and generates *magnetic field* shown by the arrows with no fill.

The RHGR is: *fingers* are in the direction of the current, *thumb* is in the direction on the magnetic field.

The acceleration of the propagating *blue shift* quantum impacts of the electron processes (the current) within the wires is perpendicular to the axis of the core.

Fig. 3.13

Figure 3.13

Ref. p.25

With reference to point 25 of the earlier section, the reason of the impact is the increased intensity of the *blue shift* quantum impact of the electron processes, the acceleration. The flow within the coils of the solenoid means revolving electron process *blue shift* impacts all around the core of the solenoid.

ef. p.15- -21

The *current* means the propagation of the *blue shift* quantum impact within the solenoid from an external source. With reference to points 15-25, the flow and the circulation generates anti-electron process *blue shift* surplus within the wires of the coils. The surplus has its quantum impact, directed on each discrete points of the inside area of the core; even in the case of an air core as well.

Once the rotation starts, the impacts, perpendicular from discrete points to discrete points until the start, become *bent*, against the direction of the rotation. It means in relative terms the direction of the current within the coils, just in the opposite direction!

Ref. p. 22-28

In relativistic terms the situation can also be explained the way that the core is the one in rotation and the current in the coils of the solenoid are at rest. With reference to points 22-28 the quantum impacts of the anti-electron processes of certain specific directions of the magnetic field of the rotating core are influencing the elementary processes of the wire of the solenoid. The developing conflict generates the flow of electron process *blue shift* impact (current) within the wires of the coils.

As the rotation is one and the same all along the core, the direction of the generating current within the wire of the solenoid is one and the same, opposite to the direction of the supposed rotation of the magnetic field independently on the angle of the plain sections of the coils.

In the case of *relativistic approach* and supposed that there is *no electricity* flow within the wires of the solenoid and the core with magnetic field is the one, which is rotating, the effect fully represents the case, just from opposite ends:

There is anti-electron process communication between the elementary processes of the solenoid and the magnetic core even without rotation.

The rotation is increasing this impact, as the radial acceleration is increasing the intensity of the elementary processes of the core.

the standard direction of anti-electron process *blue shift* impacts

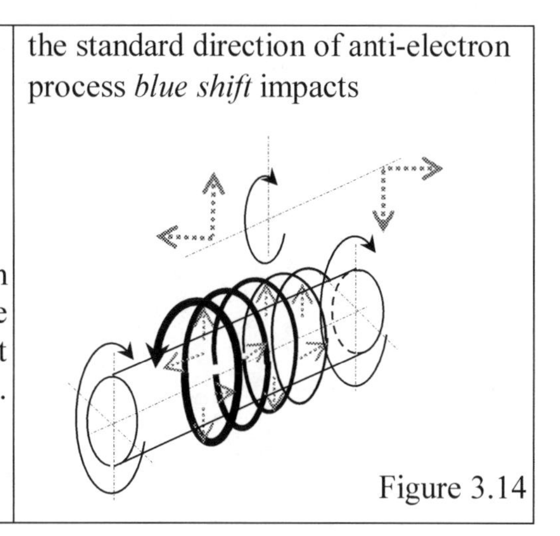

Figure 3.14

Fig. 3.14

The direction of the current relative to the rotation is clear: It is opposite to the direction of the rotation of the core.

But what is the direction of the magnetic field (within the core, supposed to be in rotation) relative to the developing current? What the relation of the two is?

For answering these questions, the complexity of the communication of all acting magnetic fields shall be addressed first.

There are three factors influencing the direction of the magnetic fields and impacts all above the surface of the *Earth*:

(1) the magnetic field of the *Earth*; (2) the gravitation; and (3) the rotation of the *Earth*.

The magnetic field within the *Earth* is directed from *North* to *South*; above the *Earth* surface is directed from *South* to *North*; the quantum impact of the gravitation is acting from the surface and directed upwards. The resultant of these two equal quantum impacts is directed upwards and twisted to *North* all over above the *Earth* surface. The rotation of the *Earth* bends the resultant quantum impact in opposite direction to the tangential speed!

The quantum impact of the gravitation, the magnetic field and the rotation of the *Earth* establish a *global, three dimensional magnetic "cloud"* (GMC) all above the *Earth* surface in any highs!

All magnetic relations above the *Earth* surface are subjects to this global impact and shall correspond to this impact.

If the direction of the relative rotation of the core of the electromagnet concise with the direction of the rotation of the *Earth*, the acting impact of the magnetic field of the core and the generating current, concise with the direction of the general magnetic cloud (GMC) of the *Earth*.

The image of this *global magnetic cloud* and its direction in three views is: 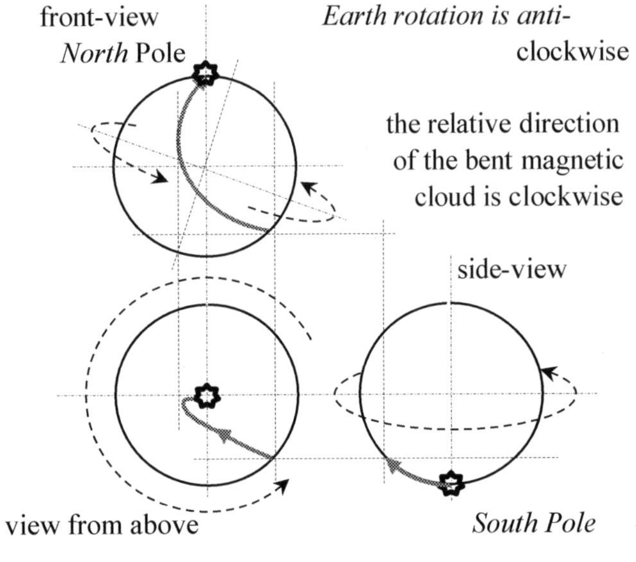 Figure 3.15	As the *Earth* is rotating and revolving around the *Sun*, and both directions are *anti-clockwise*, the all over relative global magnetic impact above the surface is directed upwards, twisted to *North* and bent <u>*clockwise*</u>, in opposite direction of the rotation-circulation of the *Earth* and corresponds to the direction of the *compass*! The intensity of the global magnetic cloud (GMC), relative to the intensity of the acting in relations magnetic fields of the cores is minimal but irrelevant; instead the *direction* is relevant.
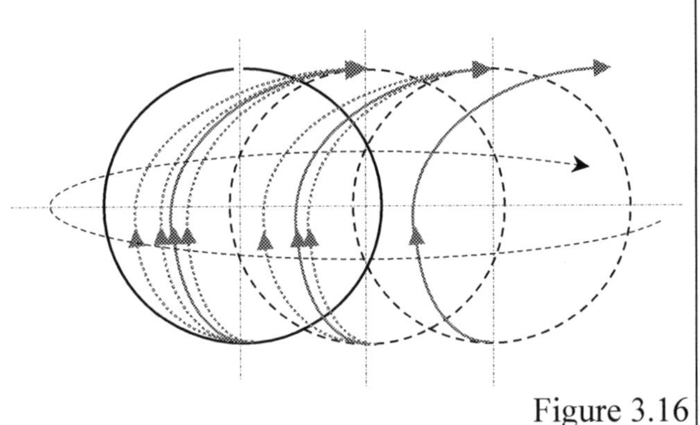 Figure 3.16	The direction of the *global magnetic cloud* (GMC) is continuous and is acting in a certain spiral; in line with the motion of the *Earth*; the circulation around the *Sun* and at the same time the rotation around its axis of rotation; the generating magnetic field is incorporating all possible three-dimensional directions of the magnetic field of the *Earth.*
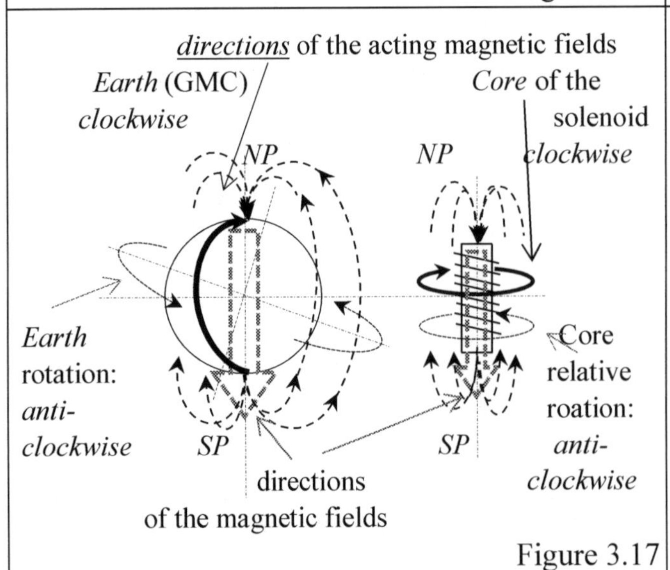 Figure 3.17	The direction of the acting magnetic field of the core, impacting the electron processes within the coils of the solenoid is opposite to the direction of the rotation of the *Earth.* If the relative rotation of the core is anti-clockwise, the acting impact (and the generating current) is *clockwise,* the direction of the magnetic field of the core concise with the direction of the GMC.

Fig. 3.15

Fig. 3.16

Fig. 3.17

In the case of the anti-clockwise direction of the relative rotation of the core of the solenoid or the clockwise direction of the current, the acting impact of the magnetic field of the core and the GMC and the direction of the magnetic impact of the core of the *Earth* are in full harmony.

This statement is based on the followings:
Once (1) the direction of the current is clockwise or the rotation of the core is anti-clockwise and (2) the rotation of the *Earth* is anti-clockwise or the relative circulation of the *Sun* is clockwise, *the direction of the magnetic field of the core* of the solenoid concise with the direction of the magnetic field of the *Earth.* As the rotation of the *Earth* has its fixed direction and the direction of the current is also fixed, as it is given in this certain case, the third component of the relation, the direction of the magnetic field has already been determined!
In a three-component-equal-relation two from the three always determines the third.

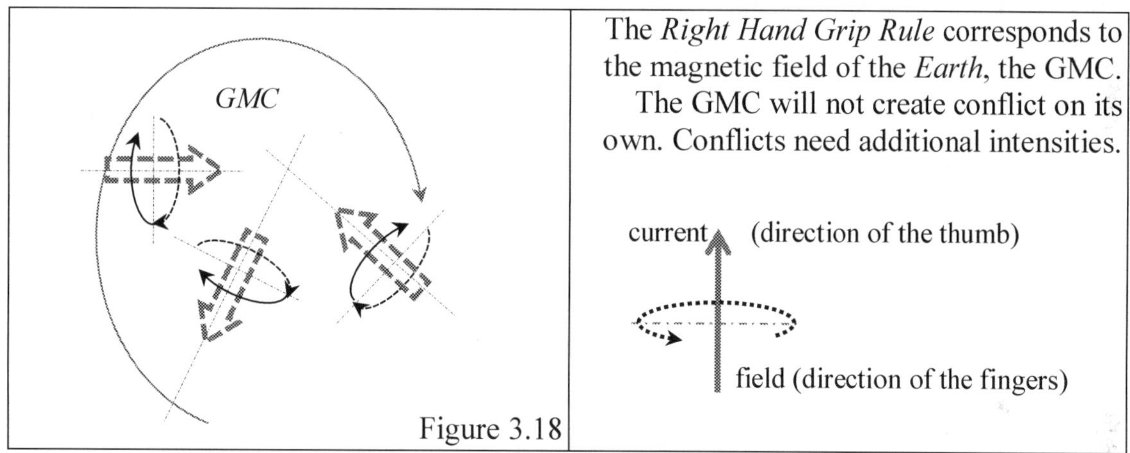

Figure 3.18

Fig. 3.18

In other formulation:
The resultant magnetic impact of the *Earth* does permit only similar relations. Otherwise the communication of the magnetic impacts of the *Earth* and other electromagnets would result in conflict.

The motion of the GMC and the rotation of the magnetic fields (the core of the solenoids) are expressed in relativistic terms. Relativistic relations however – if all conditions are taken relativistic way – are valid and represent the real case.

The *RHGR* in the real practice works in simple way: the direction of the magnetic field of the core does only concise with the direction of the GMC, if the direction of the current concise with the direction of the rotation of the *Earth.* The switch of electricity to the electromagnet (with certain direction of the wiring of the solenoid, as *input*) initiates the start of the communication with the global magnetic cloud, with immediate establishment of the direction of the acting magnetic field. The quantum impact gives the command, forming the direction of the electromagnetic field of the core. GMC assures the harmony and as automatic and natural response, excludes the other option.

S. 3.4.2

4.2. The Right Hand Rule

With reference to the picture on Figure 3.4, RHR is for the demonstration

- of the generation of electricity within a wire in motion under electromagnetic impact; and for its inverse
- of the motion of a wire with current under the impact of electromagnetic field.

the *thumb* is in the direction of the force impact (the motion); the *trigger-finger* shows the direction of magnetic field; the *middle-finger* the direction of the current within the wire.

As of the earlier section, the harmony with the direction of the GMC is the basis of the magnetic relations.
The magnetic field in relative rotation generates anti-electron process quantum impacts, a density of certain value.
The movement of the wire with electricity within a magnetic field is result of the conflict of different magnetic impacts acting in parallel.

What the direction of the current within a wire in motion is? What the direction of the acting force effect is? What the relation of the two is?

The magnetic field is given as input, with arbitrary direction within the global magnetic cloud.

It is supposed, that the magnetic field between the *Northern* and the *Southern* poles of the magnet is in rotation: either anti-clockwise, in line with the rotation of the *Earth* or in clockwise, against it.

The loop, between the poles of the magnet is at rest.

As the direction of the relative motion of the GMC is *clockwise*, in the case of clockwise direction, it concise and harmonise with the direction of the motion of the GMC.

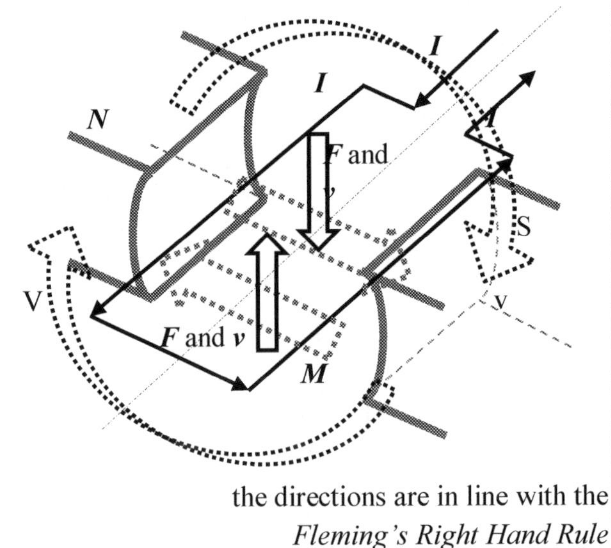

the directions are in line with the *Fleming's Right Hand Rule*

Fig. 3.19

Figure 3.19

This relation on Figure 3.19 above shows the generation of the *alternate current*. The magnetic field is constant without any change and motion; the wire between the magnetic ends is rotating, which results in permanent change of the direction of the force impact; the current is the one which follows the change of the direction of the rotation (and the force impact).

The rotation of the direction of the magnetic field initiates current within the loop in line with the *Right Hand Grip Rule* (RHGR) in each its rotating positions.

3.5
Complex magnetic fields

S. 3.5

There are two solenoids on Figure 3.20. One is built into the other. They are electrically connected in cascade format. The directions of the current in the solenoids are opposite to each other.

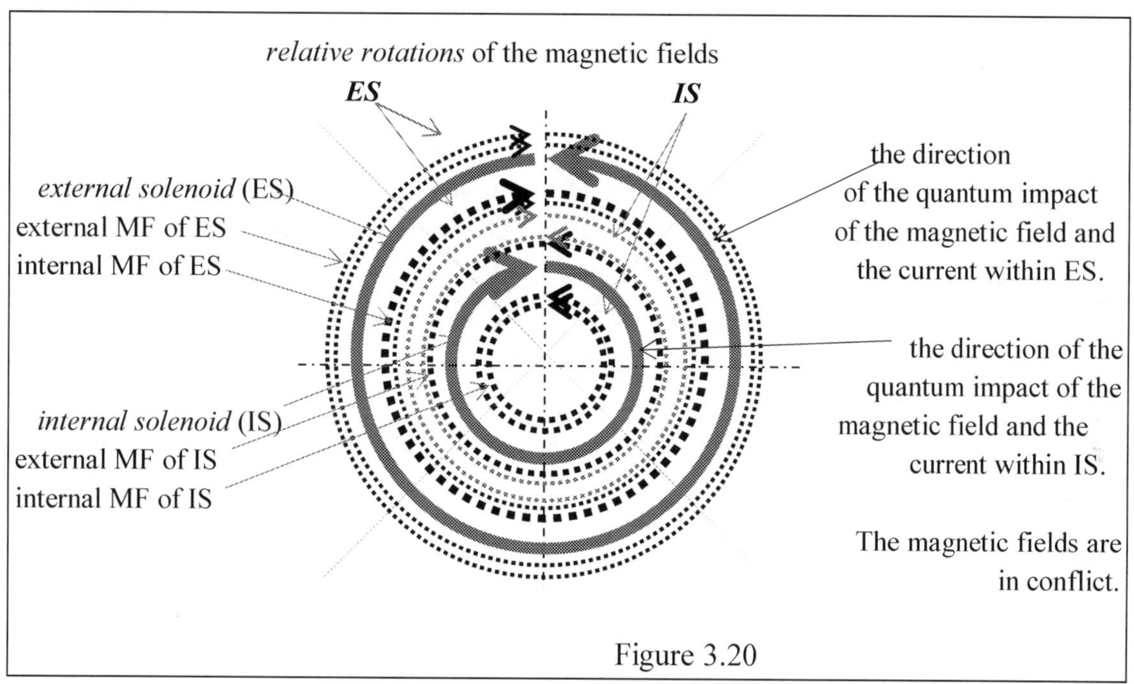

Figure 3.20

Fig. 3.20

With reference to the earlier sections, the magnetic field is the quantum impact of the surplus of the anti-electron processes. The surplus is given off by the elementary processes. In the case of a solenoid, spooled by coils all around, the magnetic impact is given off in both directions: towards the internal and the external areas. The difference is, that giving off it inward, the surplus meets similar impacts. The density of the impact is increasing. The only places, to leave the area of the internal conflict are the surfaces at the two ends of the air core, with attractive effect, as discussed earlier. Leaving in the directions towards the external areas is easier, but with the decrease of the density of the impact.

The case, with solenoids with air core, built into each other is special. The magnetic field of the internal solenoid is acting within the magnetic field of the external solenoid, both with air core. The inward impact of the external solenoid is meeting the outward impact of the internal solenoid.

The radial directions of the magnetic fields are opposite to each other. The values of the current flowing through both solenoids connected in cascade format are obviously of equal. The directions of the wirings of the solenoids are also opposite to each other. The relations of the directions of the magnetic fields in the solenoids correspond in both cases to the RHGR of the Global Magnetic Cloud.

There are numbers of specifics to be taken into account during the assessment:

1.

For the fluent propagation of the *blue shift* quantum impacts of the electron process in the wires of the solenoids *the tangential speed* of the current within the loops shall be *equal,* independently of the number of the coils!

The equal values of the tangential speeds at different radiuses means different angular speed values of the relative rotations of the magnetic fields. The less the radius of the solenoid is, the higher is the angular speed of the relative rotation.
This means the angular speed of the rotation of the magnetic field of the internal solenoid is always higher than of the external solenoid. The radius is the one defining the angular speed of the rotation. The equal tangential speed values in the coils of the solenoids of different sizes need different driving forces of the rotation of the magnetic fields. For having equal current values (equal tangential speed values) in the solenoids, the external solenoid needs less driving force. This means in the opposite, not relativistic way that the same value of the current (with less voltage drop and less value of resistance) generates a magnetic field of less intensity.

2.

With reference to the *Biot-Savar* formula, the strength (the intensity) of the magnetic field is the function of the value of the current and the radius of the solenoid:

3A1 $$B = \frac{\mu_o I}{4\pi r^2} l\,; \quad \text{where } \frac{\mu_o}{\pi} = const, \qquad \text{meaning: } f(B_x) = f\left(\frac{I_x}{r_x^2}\right);$$

The electricity flows and the lengths are the same, therefore:

3A2 $$f(I \cdot l) = f(r_i^2 |B_i|) = f(r_e^2 |B_e|)\,;$$

In line with para 1, at equal lengths and current values the magnetic field of the internal solenoid is stronger

3A3 $$|B_i| = \frac{r_e^2}{r_i^2} |B_e|$$

The stronger magnetic field in relativistic terms means, stronger rotating torque value. In fact, this is the case, since for the generation of the same tangential speed (the equal values of the current) the density (the “mass”) of the magnetic field at shorter radius shall have higher angular speed. The higher value of the angular speed corresponds to the higher number of the loops. In line with 3B3, later in the section, the intensity of the magnetic field is proportional to the number of the loops (the angular speed) and the cross-section of each of the loops (in fact the volume, the quadrate of the radius) of the solenoids (at equal lengths):

3A4 $$\tau_x = f(B_x) = f(n_x r_x^2);$$

Equal values of the magnetic fields need more number of loops at shorter radius.

3.

The directions of the currents define the directions of the magnetic fields (the directions of the relative rotations).

Ref. S. 3.1

If the direction of the external field is taken as positive and attracting, the direction of the internal magnetic field is negative but, with reference to Section 3.1, is also attracting.

The rotating magnetic fields have their own torque values.

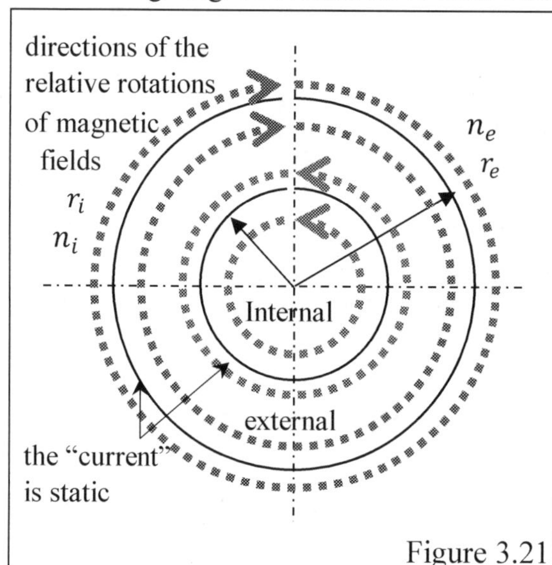

Figure 3.21

The definition of the torque value is:

$$\tau_x = F_x r_x; \quad \text{and } F_x = m_x \frac{dv}{dt} \rightarrow$$

$$B_x = \int m_x \omega_x \frac{dr_x}{dt}; \qquad \text{3B1}$$

The *mass* of the magnetic field means its "density", function of the volume and the length of the solenoid.

$$m_x = f(r_x^2 \pi \cdot l); \qquad \text{3B2}$$

This way: $m_x = f(r_x^2)$

$$B_x = f(n_x r_x^2); \quad \text{as } n_x = \omega_x \qquad \text{3B3}$$

and $n_i = \frac{r_e^2}{r_i^2} n_e$ 3B4

Fig. 3.21

In line with the RHGR, the directions of the two magnetic fields are opposite to each other. While the attractions at both ends with reference to points 44-48, 52-56 of Section 3.1 in normal conditions are still valid, the conflict of the two magnetic fields in certain conditions results in the unusual *repulsion*, instead of the expected attraction.

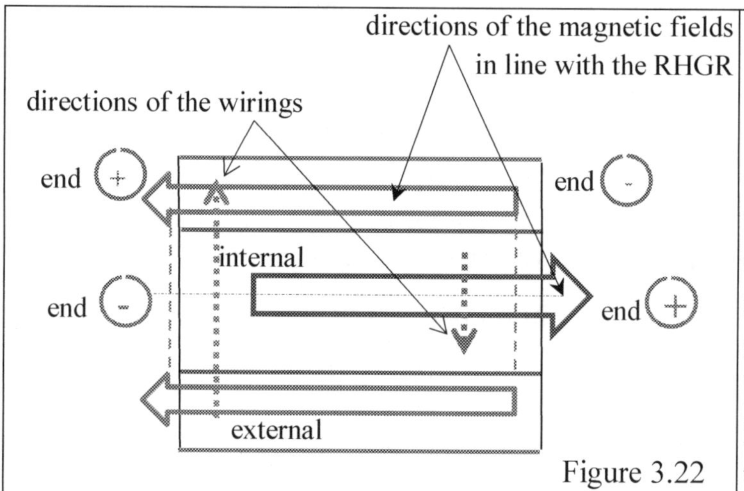

Figure 3.22

The key of the conflict are the continuity of the flow of the current.

Once the strength (torque) of the magnetic field of the external solenoid is higher than the internal one, meaning $|B_e| > |B_i|$ or $|\tau_e| > |\tau_i|$ the RHGR for the internal magnetic field is not valid anymore.

Fig. 3.22

The effects of the magnetic field of the internal solenoid are opposite.

At the PLUS END of the internal solenoid, the result is: $(-) - (+) = (-)$, as the direction of the wiring does not correspond to the RHGR. Instead of attraction, it is REPULSION! 3B5

At the MINUS END of the internal solenoid, the result is: $(+) - (-) = (+)$, but the direction at this end should be minus. The result is instead of attractions: REPULSION! 3B6

The impact of the external solenoid breaks the power of the magnetic field of the internal solenoid.

The impact works and valid in the opposite way as well. In the case the strength of the internal magnetic field is more than the external one, the place of the generation of the repulsion force is the concentric sector between the external surface of the internal solenoid and the internal surface of the external solenoid.

The conflicts are about:

1. the *attraction* is the natural impact of both solenoids;
2. the directions of the current in the solenoids are opposite;
3. the parallel use of the RHGR is not possible;

Ref. 3A1 3C1 3C2

With reference to 3A1, the relation of the current and the magnetic field of the external solenoid is: $\frac{+I}{+B_e} = x_e$;

The same relation of the internal solenoid is: $\frac{-I}{+B_i} = -x_i$;

the $(+B)$ sign marks the *attraction* in both cases;
the $(+I)$; and the $(-I)$; mark the directions of the current within the solenoids, which are opposite;
The relation of a real current value and the real magnetic impact cannot be of negative sign!

The only way for resolving the conflict is if the directions of the magnetic impacts of the solenoids were also opposite! The relation of one of the solenoid, therefore is:

3C3 $\frac{-I}{-\Delta B_i} = x_{\Delta i}$; this is the relation, which resolves conflict of 3C1 and 3C2

in this case $(-\Delta B_i)$ shall mark the opposite impact of the attraction = *repulsion*! And the practice in certain circumstances proves it.

The direction of the summarised impact depends on the strengths of the acting magnetic fields, which are the function of different components to be discussed in the followings. The impacts at both ends of the internal solenoid are REPULSION!

S. 3.5.1

3.5.1 The relations and the variants of the attraction and repulsion impacts

Ref 3A2 3B4

With reference to 3A2 and 3B4, there are here three kinds of options as results of the communication of the fields:

3D1 3D2	$\frac{\lvert B_i \rvert}{\lvert B_e \rvert} = \frac{r_e^2}{r_i^2}$; $\frac{n_i}{n_e} = \frac{r_e^2}{r_i^2}$;	The quantum impacts of the magnetic fields are equal. The torque values of the rotating magnetic fields are equal. The balance results in magnetic quantum impact; the impacts at both ends of the solenoids are *neutral*. The numbers of coils obviously are different.
3D3 3D4	$\lvert B_i \rvert < \frac{r_e^2}{r_i^2} \lvert B_e \rvert$; $n_i < \frac{r_e^2}{r_i^2} n_e$;	The quantum impacts of the magnetic field and the torque value of the internal solenoid are of less value. The result is *repulsion* at the both ends of the internal solenoid.
3D5 3D6	$\lvert B_e \rvert < \frac{r_i^2}{re_e^2} \lvert B_i \rvert$; $n_e < \frac{r_i^2}{r_e^2} n_i$;	The quantum impacts of the magnetic field and the torque value of the external solenoid are of less value. The result is *repulsion* at the both ends of the concentric section of the external solenoid.

The magnetic impacts depend on the materials of the core of the magnet. In the case the materials of the cores are the same, the relation in 3D1-3D6 are valid for use. In the case the cores are of different materials, the magnetic properties of the materials shall be also taken into account. With reference to 3A1

Ref. 3A1 3D7

$$f(B_x) = f\left(\frac{\mu I_x}{r_x^2}\right); \quad \text{and} \quad \frac{B_x r_x^2}{\mu_x} = const$$

The equality of the values of the magnetic fields, as of in 3D4 would mean balance and no all-over impact at the ends of the composition. But this is the condition which has just been changed.

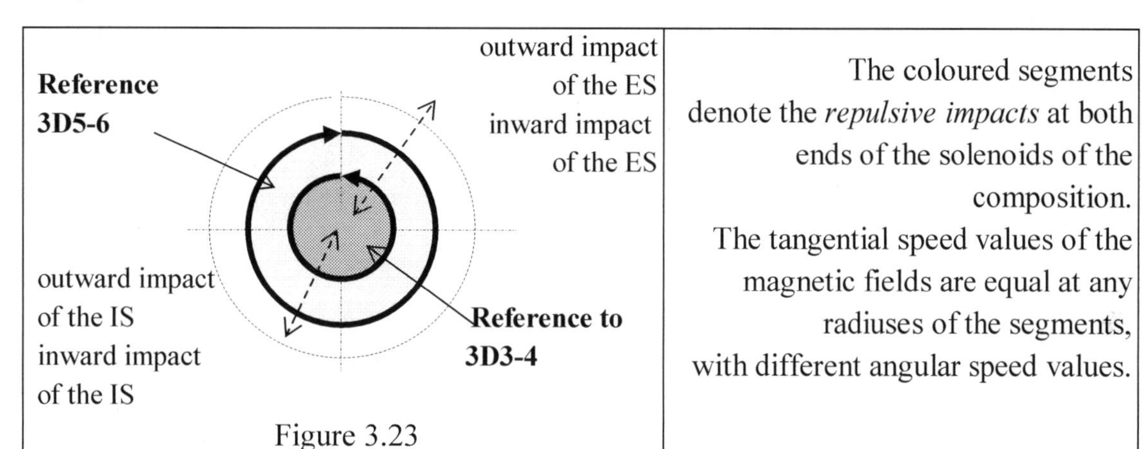

Figure 3.23

Fig. 3.23

In the case of the *repulsive* magnetic impact at specific radiuses, the number of the coils of one of the solenoids defines the number of the coils of the other solenoid. The relation of the unit values of the change defines the area of the repulsion.

3.6
The magnetic wind

S. 3.6

The repelling impact can be called, with reference to Figure 3.24 below, as *magnetic wind.*

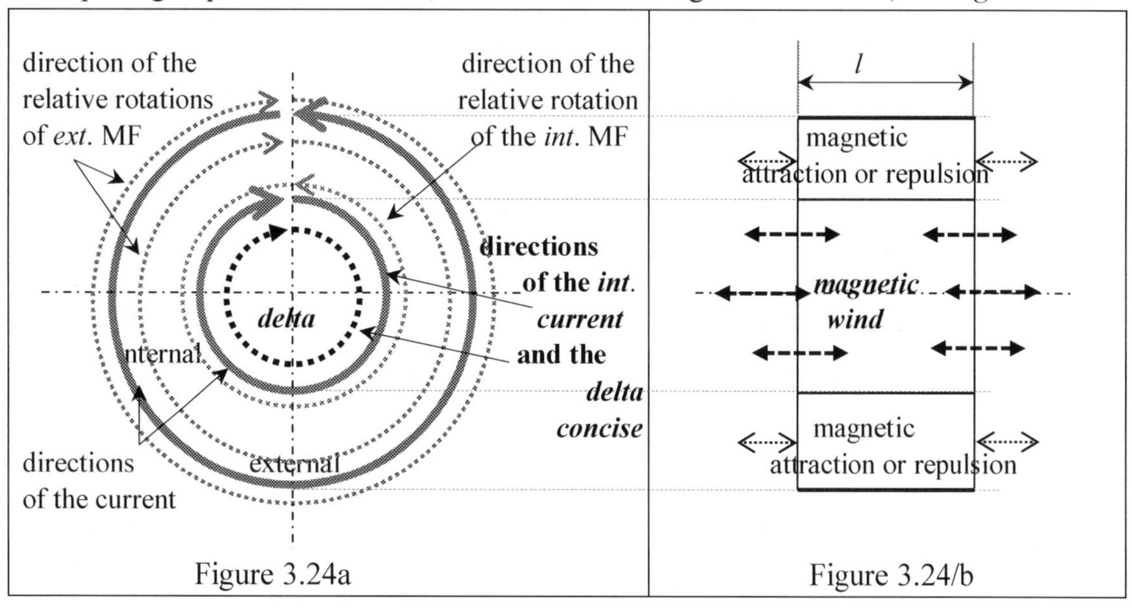

Figure 3.24a

Figure 3.24/b

Fig. 3.24

With reference to 3D3, 3D4 and 3D5, 3F9 the space between the external and the internal solenoids, or the internal space of the internal solenoid work as *attractive* or *repulsing* magnetic fields.
A composition of the electromagnets with magnetic wind needs equal length of solenoids.

The *magnetic wind* is *repulsive* at both ends of the internal solenoid!
The magnetic wind is the release of the surplus of the quantum impact of the anti-electron processes of the magnetic fields.

In the case of three or more solenoids built into each other the magnetic blow can only be detected in front of the internal solenoids. And – as the experiences prove it – it is of relatively weak value. The reason is that the developing step by step surpluses between the solenoids are resolved (or covered) by the solenoids in the internal space in the direction towards the centre. Just the developing surplus in the last solenoid is detectable and released.

S. 3.7

3.7
Magnetic wind for transport

There are the data of the magnetic impacts on Figure 3.25, measured during the experiment with the electromagnets, connected in cascade format, with attracting at one end and with repulsing impacts on the other. The transporting function needs different numbers of the coils at the two ends of the solenoids. The increased number of the coils of the external solenoid at one end generates repulsive impact, while the other end is still an attracting one.

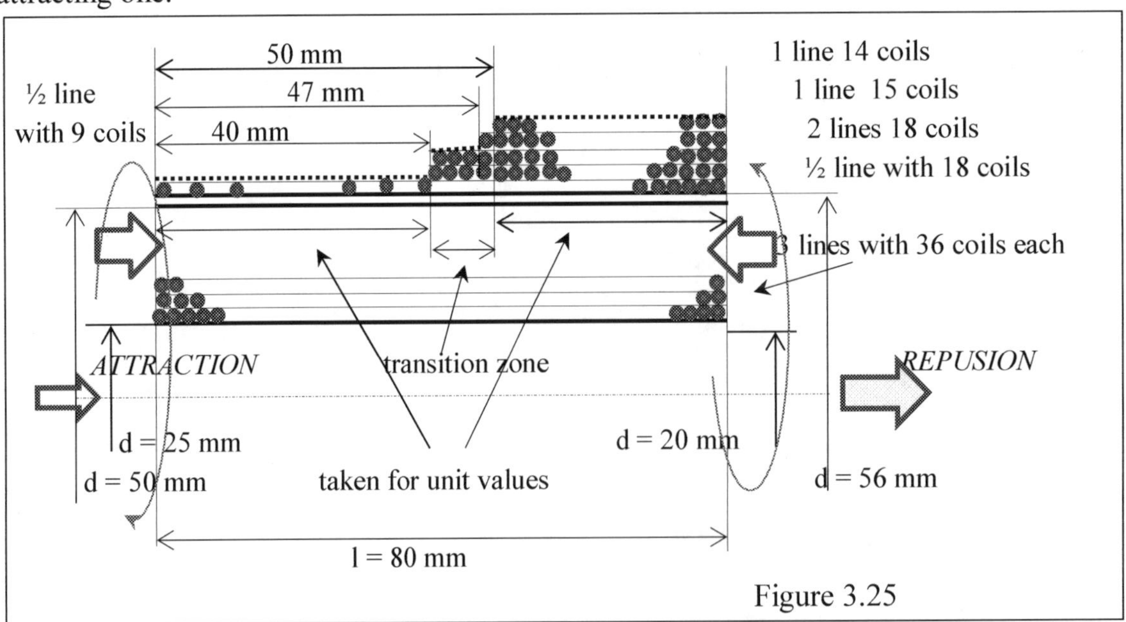

Fig. 3.25

Figure 3.25

There is the scheme of two solenoids built into each other on the Figure 3.25.
The power of the attraction and the repulsion impacts depends on the number of the coils at the two ends. This composition of electromagnets was used as the *quantum drive* in the experiment of the acceleration of the *Hydrogen* process, discussed in details in Section 9.

Relations at the ***attracting*** end in unit values	$\frac{r_e}{r_i} = 2;\quad n_i > \frac{r_e^2}{r_i^2} n_e$; proving the ***attraction***, as normal function of the composition of the electromagnet	3E1
Relations at the ***repulsing*** end also in unit values	$\frac{r_e}{r_i} = 2;\quad n_i < \frac{r_e^2}{r_i^2} n_e$; proving the ***repulsion*** impact	3E2

The diagrams below show the numbers of the rows and the coils of the quantum drive.

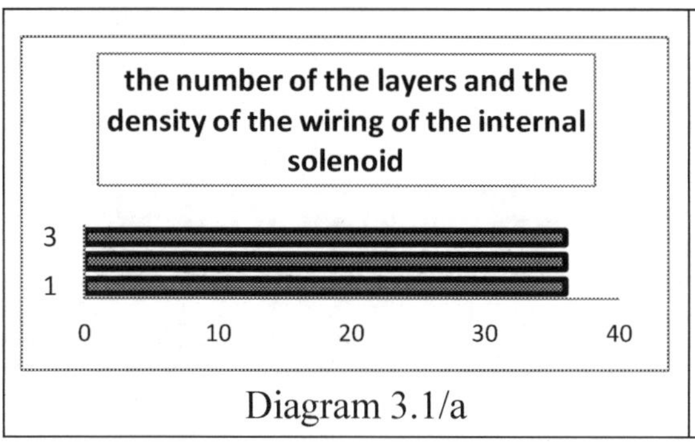

Diagram 3.1/a

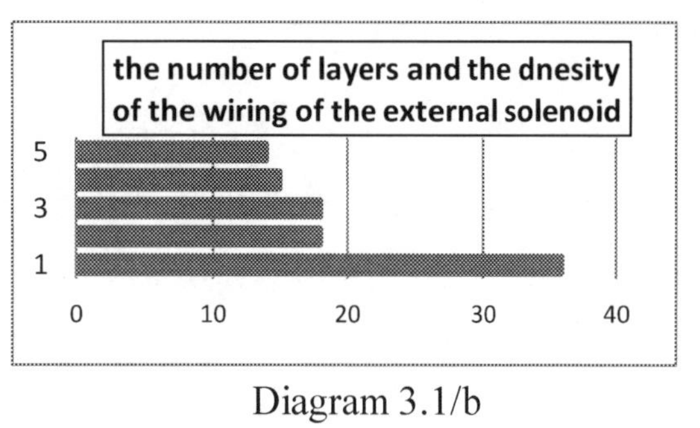

Diagram 3.1/b

Diag. 3.1

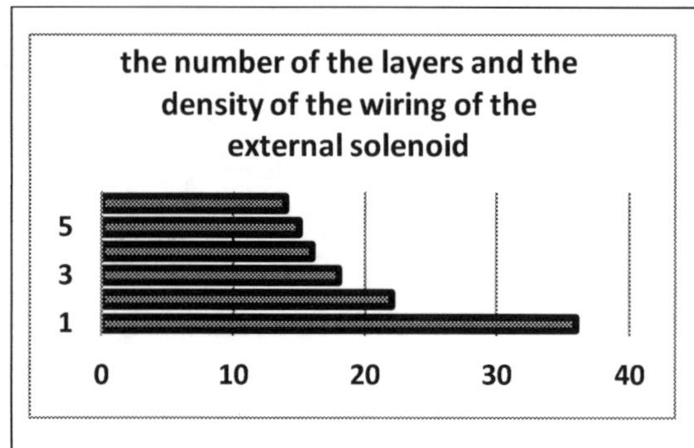

Diagram 3.2

Another experiment proves that the increase of the impact of the number of the coils of the external solenoid is significant.

$$n_e = \frac{r_i^2}{r_e^2} n_i\ ;$$

The number of the coils of the external solenoid with attraction at one end and the repulsion of the other in certain sequence is ***96*** as the diagrams show it.

Diag. 3.2

The number of the coils on the 80 mm length of the solenoid in the 6 lines around the external diameter	1	2	3	4	5	6
	36	***22***	***18***	***16***	***15***	***14***

In the next example/experiment the cross-sections of the wires are different. The cross-section of the wire with $d_{ew} = 1$ mm is 4 times more than that of the wire with $d_{iw} = 0.5$ mm.

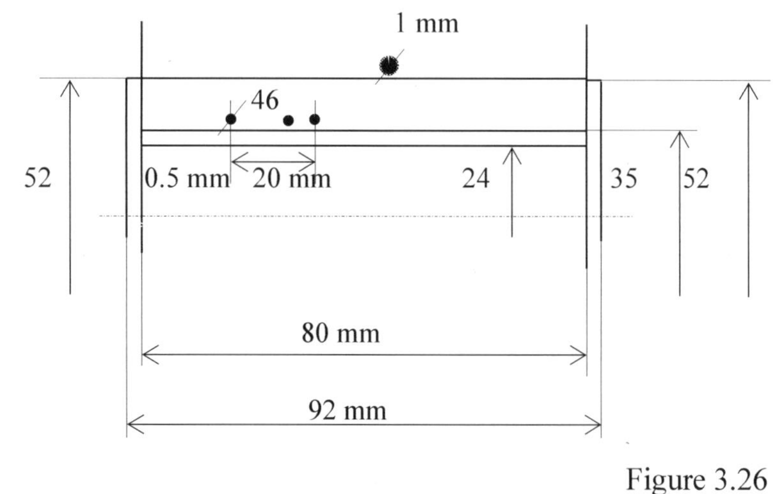

Figure 3.26

The diameter of the internal and the external solenoids are
$d_i = 24$ mm
$d_e = 52$ mm

$n_{iw} = 566$ in 4 lines

For having balance the number of the coils of the external solenoid shall be: $n_{ew} = 256$
The acting repulse and the attraction however need ≈ 400.

Fig. 3.26

3F1 With reference to Figure 3.26, the values of the current in the wires are equal $I_{iw} = I_{ew}$,

3F2 But $\frac{A_{iw}}{A_{ew}} = \frac{1}{4}$ the densities of the flows are different, therefore $4\rho_{e(flow)} = \rho_{i(flow)}$

Ref. 3B4 3D4 With reference to 3B4 and 3D4, for the magnetic fields, the wiring of the external solenoid shall be: $n_e = 256$, as of

3F3

$$n_i = \frac{r_e^2}{r_i^2} n_e; \quad \frac{r_e^2}{r_i^2} = 4.7; \quad |B_i| = \frac{r_e^2}{r_i^2} |B_e|; \text{ and } |B_i| = 4|B_e|$$

3F4 The length of the wires: $\frac{l_{iw}}{l_{ew}} = \frac{53.3}{75.5} = 0.7$

3F5 $R = \rho \frac{l}{A}; \quad \rho_{ew} = \frac{1}{4}\rho_{iw}; \quad R_x = f\left(\rho_{x(flow)} \frac{l_x}{A_x}\right); \quad I_{iw} = I_{ew}; \quad U = I_x R_x; \quad U_x = f(R_x);$

3F6 $U_x = f\left(\rho_{x(flow)} \frac{l_x}{A_x}\right); \quad \frac{U_{iw}}{U_{ew}} = 11.3 = \frac{R_{iw}}{R_{ew}}; \quad \frac{U_{ew}}{U_{iw}} = 0.09;$

3F7 In the case of $U_{sum} = 25\ V$, the voltage drops within the wires are

$U_{iw} = 22.8\ V$; and $U_{ew} = 2.2$ V;

4
Experiments with electromagnets
The experimental proof of the statements in Chapter 3
Magnetic Wind

The electromagnets ***EM1*** and ***EM2*** may have different formats and connections.

If they are connected in opposite directions =
= the results is no magnetic fields in *EM1 and EM2*

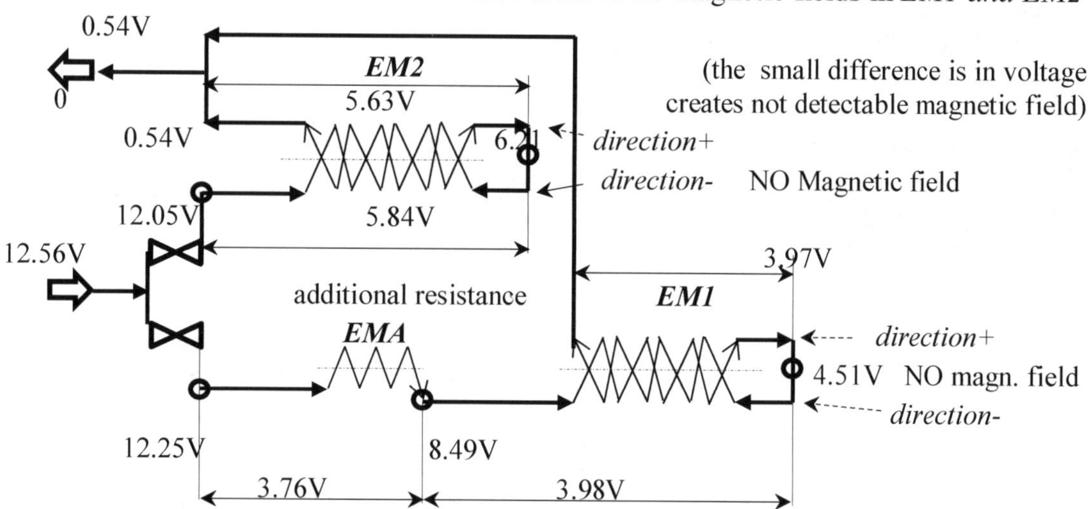

There is an additional electromagnet (*EMA*) connected into the cycle in parallel, just for having the relation in a lower voltage value and having in this way a magnetic field of less value as well.

Figure 4.1

Fig. 4.1

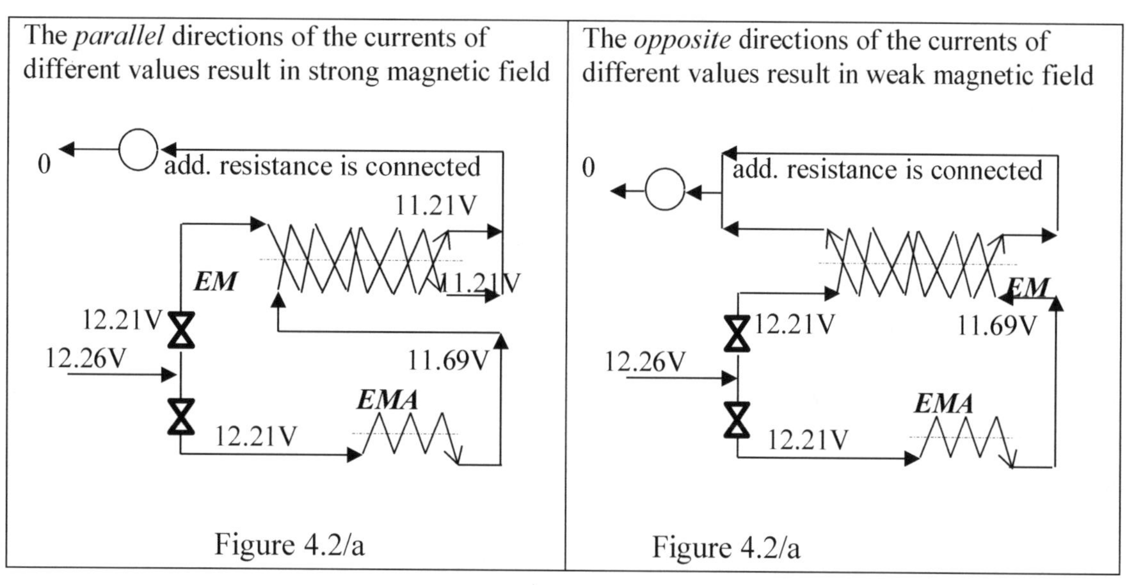

Figure 4.2/a

Figure 4.2/a

Fig. 4.2

The wirings in Figures 4.1 and 4.2 were made on one and the same solenoid!

The parallel directions are strengthening; the opposite directions are weakening the magnetic field.
Without the complex composition of the solenoids and being separated, are the values of the currents equal or different, are the directions of the same or the opposite, the developing magnetic fields have always *attracting* effects; and never a *repelling* one.

Ref. Ch.3 Fig. 29, 32

With reference to Figures 3.20, 3.21 and 3.22 of Chapter 3, the impact is different if the solenoids are built within each other.

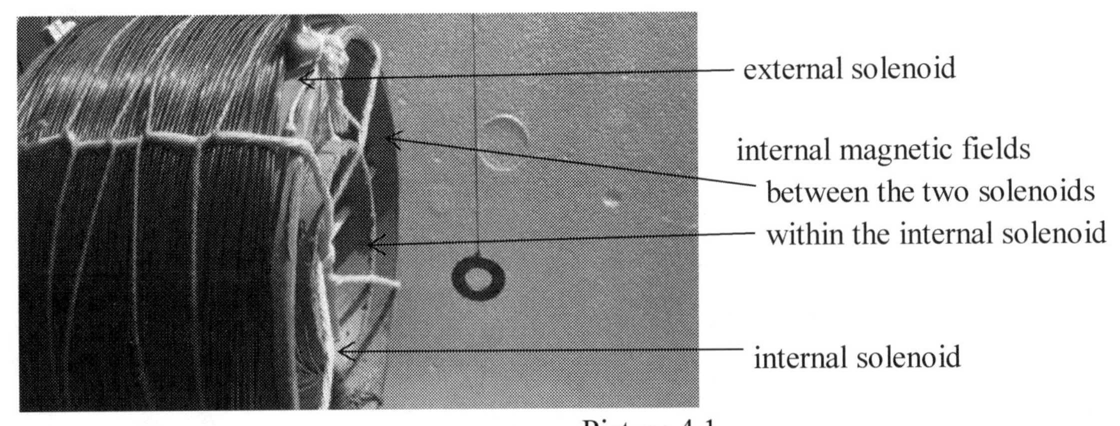

Pic. 4.1

Picture 4.1

attracting impact

repelling impact

attracting impact

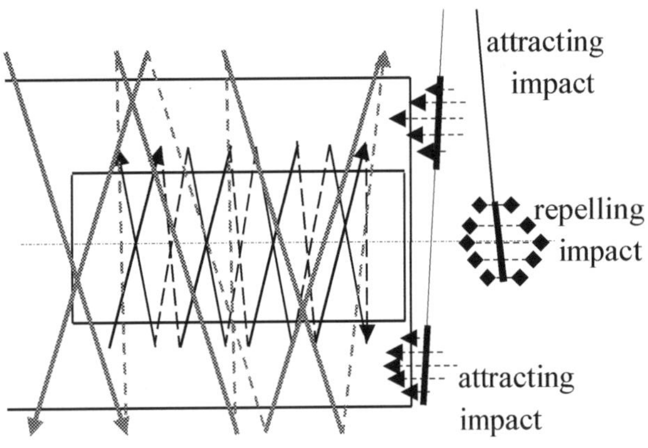

Figure 4.3

It seems that the magnetic field of the internal solenoid does not follow the GMC rule.

The magnetic field of the internal solenoid becomes impacted by the magnetic field of the external solenoid. The current is the same, but the flow is of the opposite direction.

The magnetic field of the external solenoid, around the internal one has attracting impact!

Fig. 4.3

There is a *repelling* impact in front of the internal solenoid.
The external solenoid is attracting the carbon steel target, but the attraction is limited to the area/space between and in front of the external and the internal solenoids, as Figure 4.3 above shows it.

In the case of the composition of three solenoids, built into each other's magnetic field, the two internal solenoids have *repelling* impacts. The external solenoid has *attraction* in the area/space between and in front of the external and the internal solenoids.
As the further experiments will give the proofs, the most efficient way for the generation of a repelling impact is the composition just of two solenoids.

The current flows through the solenoids connected in cascade format and generates magnetic field with specific impact in each.	The external solenoid: with direction: + → − The solenoid in the middle with direction: − → + The internal solenoid with direction: + → −

The magnetic impact was tested in 3 different voltages of the load, as demonstrated below in Figure 4.3 and Diagram 4.1

Picture 4.2

Pic. 4.2

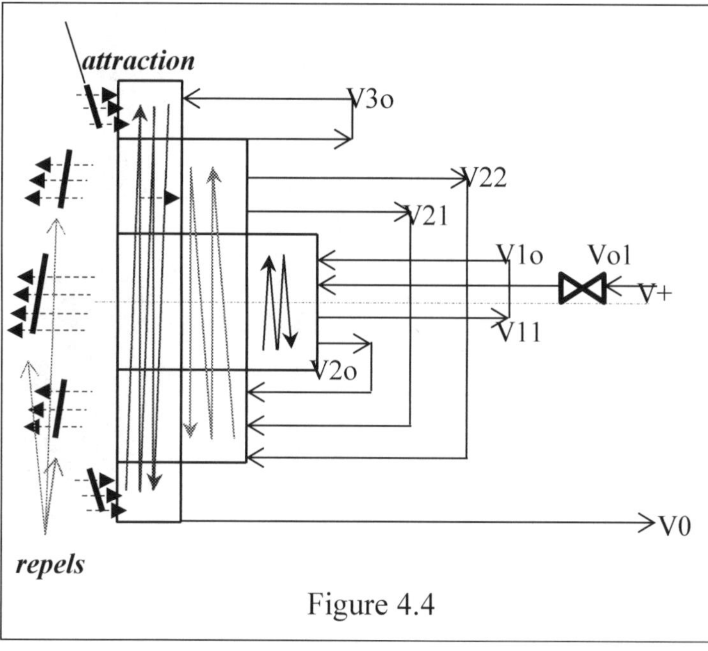

Figure 4.4

The voltages at the different points of the connections are:

	V+	12.7	25.0	37.6
1	Vol	12.4	24.3	35.8
2	V10	12.1	24.0	35.5
3	V11	11.7	23.2	32.4
4	V20	10.2	20.1	27.0
5	V21	9.3	18.3	23.2
6	V22	7.7	15.0	21.0
7	V30	5.83	11.4	16.8
8	V0	0	0	0

See the Diagram 4.1 below

Table 4.1

Fig. 4.4 Tabl. 4.1

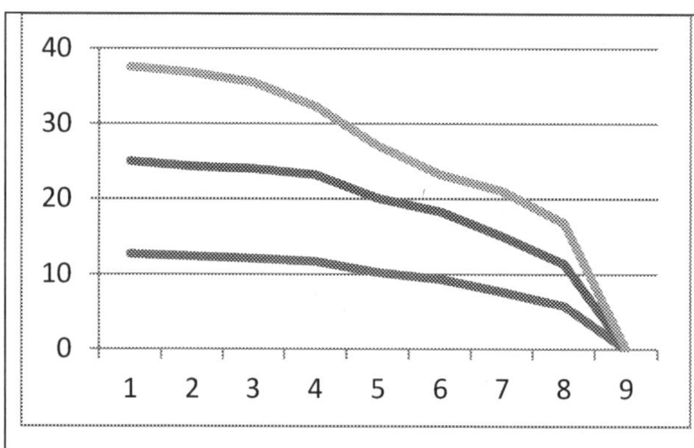

The higher the voltage was the more was the repelling and the attracting powers of the composition.

Diagram 4.1

Diag. 4.1

Vertical direction

The arrows on the picture show the repelling power of the magnetic composition.

The repulsion is not a dynamic sway! The away status is permanent.

For the better view of the repelling impact, the wire, holding the ring is strengthened on the picture.

Pic. 4.2

Picture 4.2

In the case of the composition of 5 electromagnets built into each other's magnetic fields and connected also in concentric cascade format, the distance of the repel becomes longer.

The strongest amplitude of the repulsion was obviously at the moment of the inrush current. But the repulsion was permanent.

With the increase of the initial distance of the ring at rest from the composition, the repulsion becomes less and less. (See Diagram 4.2 below)

The distance of the ring at rest from the acting surface of the EM

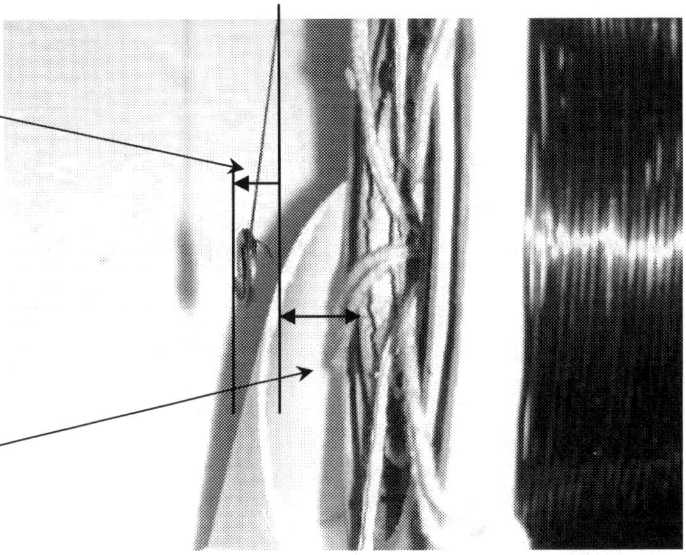

Pic. 4.3

Picture 4.3

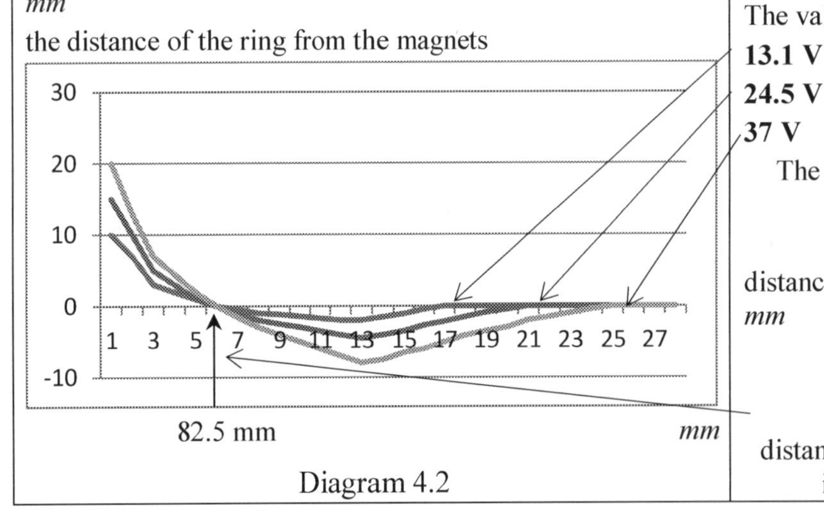

The values of the voltage are:

13.1 V

24.5 V

37 V

The total distance of the test is 150 mm. 1 unit = 15 mm

distance from the magnets *mm*

the *inflexions* are at equal distances from the composition, independent of the voltage

Diag. 4.2

Diagram 4.2

The equal distances (82.5 mm) of the *inflexions* in all three cases with different voltage values prove that each composition (the wiring and the geometry) has its unique characteristics, function of the voltage.	$a = f(U)$; and therefore $\frac{f(U)}{U} = x = const \ \frac{[m \cdot V]}{[V]}$

4A1

There are *positive* (repulsion) and *negative* (attraction) values seen in Diagram 4.2. The reason of the inflexions is that the internal electromagnets in these unique cases are of less capacity. Therefore the attracting impact of the electromagnet(s) on the periphery after a while becomes dominant with the growth of the distance more and more. The highest attracting impact at equal distance obviously belongs to the highest voltage.

In order to ensure the balanced relation and the equal distances of the magnetic impacts, the lengths of the electromagnets of the composition shall be different, deceasing towards the periphery. At different lengths of the EM however the repelling impact is far not significant. This means, the lengths of the EM shall be equal!

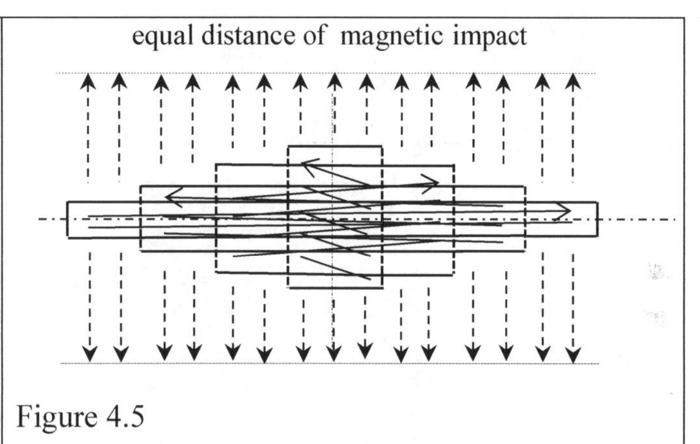

Figure 4.5

Fig. 4.5

Massive repelling impact was detected in the case of 2 solenoids, built into each other's magnetic fields with the current flowing in the solenoids in opposite directions.

For the generation of sufficient repelling magnetic impact, the composition needs certain length.

Picture 4.4

Pic. 4.4

The repelling effect of the composition of the solenoids – built into each other, with the opposite directions of the current, with increased difference in the diameters but with short horizontal length – is weak.

The external solenoid eats off the impact of the internal solenoid.

Picture 4.5

Pic. 4.5

The repelling impact (*magnetic blow*) – with solenoids, diameters of the same size as at the Picture 4.5 – is of more power, if the acting lengths of the solenoids are longer, as it is shown on Picture 4.6.

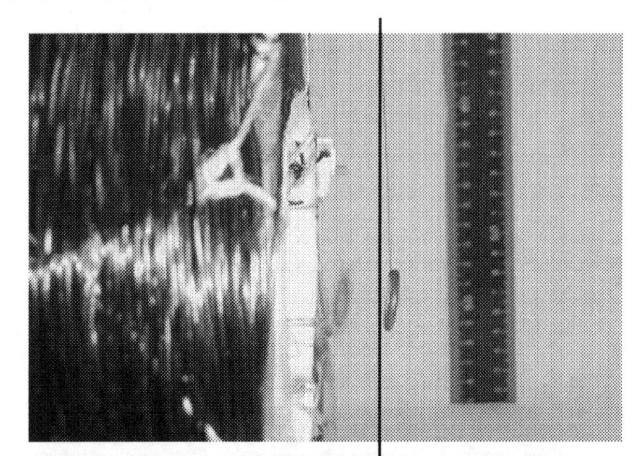

The larger the diameter of the EM is, the more powerful is the blowing impact.

Pic. 4.6

Picture 4.6

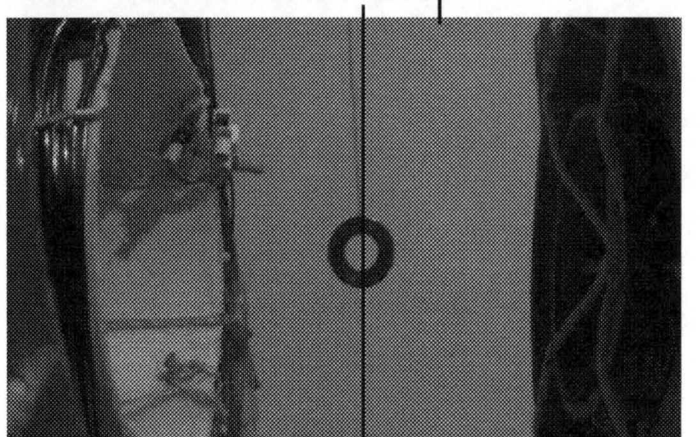

The compositions of the solenoids facing each other are confronting and acting in similar ways. The solenoid with higher capacity has the dominant blow impact.

While both solenoids generate magnetic blow, the solenoid on the left hand side has the dominance of the repulsing impact.

Pic. 4.7

Picture 4.7

As it is explained in Section 3.5, the intensity of the repulsion depends on the number of the coils of the solenoids. There was an experiment made with an internal solenoid, with one and the same number of the coils (363 turns) in all measurements, while the number of the coils of the external solenoid was changed step by step from 252 to 553.

The following tables show the measured results at different distances and windings.

The distance of the subject from the end of the centre of the internal solenoid: *1.5 cm*

Table 4.1

Impacting a large spacer ring

grade	coils	position
1	252	*attracted*
2	412	*repel* of 2.5 cm
3	439	*repel* of 2.0 cm
4	553	*repel* of 1.5 cm

Table 4.1/a

Impacting a small spring spacer ring

grade	coils	position
1	252	slight sway with *attraction* towards the internal solenoid, position: 1.4 cm
2	412	less sway, but 1.8 cm *repel*
3	439	harder sways and 1.7 cm repel
4	553	rotation, *no movement*

Table 4.1/b

As it was explained in Section 3.5, with the growth of the number of the coils of the external solenoid the *repulsion* was strengthening. The *attracting* and *repelling* impacts obviously depend on the weight of the subjects as well. The push away impact of the subjects with more weights (see Table 4.2) was of less intensity, even at shorter distance.

The distance of the impact from the centre of the internal solenoid is at *10 mm*.

Impacting a large spacer ring

grade	coils	position
1	252	attracted
2	412	repel up to 1.8 cm
3	439	repel up to 1.5 cm
4	553	repel up to 1.1 cm

Table 4.2/a

Impacting a large plastic ring

grade	coils	position
1	252	slight swinging motion in all positions
2	412	
3	439	
4	553	

Table 4.2/b

Table 4.2

There are here a YouTube videos for demonstrating the repelling impact:
https://www.youtube.com/watch?v=PI-PSC5y_NI

There are here two moments on the pictures from the video above for illustrating the push force on a simple wide spacer shim

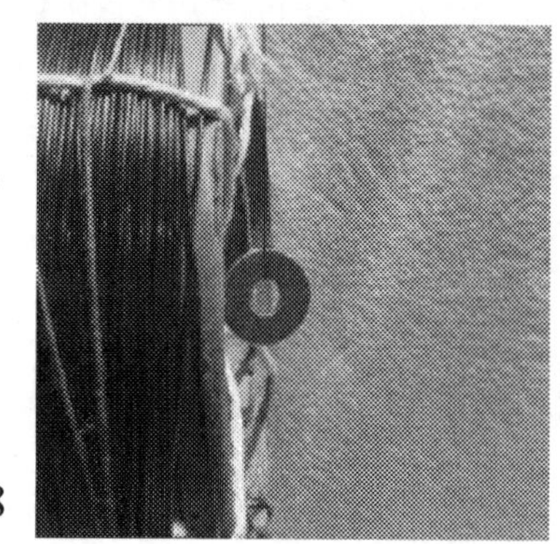
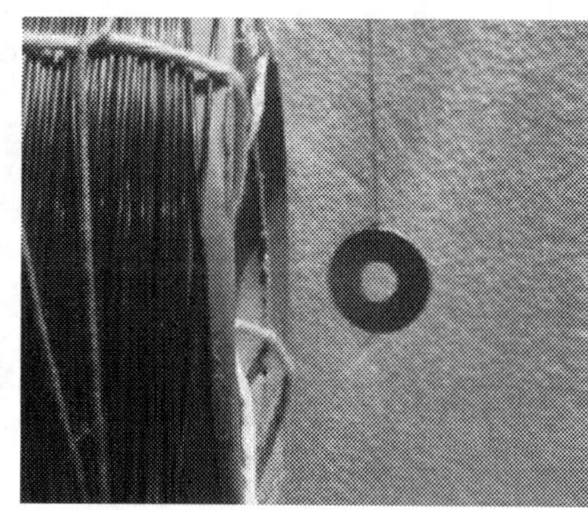

Picture 4.8

Pic. 4.8

The next YouTube video shows the power of the magnetic wind capable for the rotation of a slim CD:
https://www.youtube.com/watch?v=UZQdTAABkWU

There are here three moments from the video:

Picture 4.9

Pic. 4.9

The Magnetic Wind has its electromagnetic impact in vertical position as well.

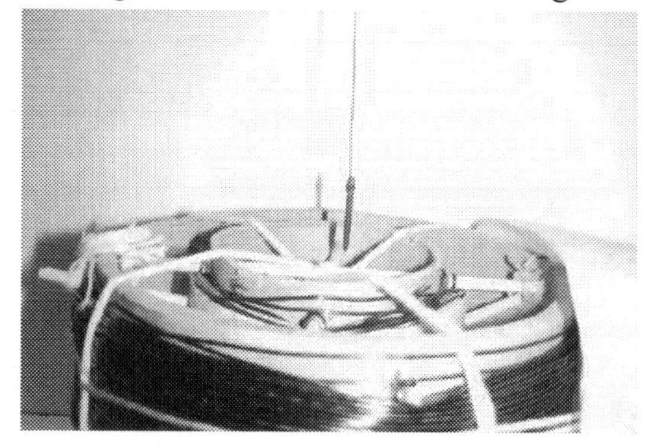

This is the same composition as in Picture 4.1, just turned into vertical position.
With no voltage in the coils the nail indicator was not impacted. It hangs above the centre of the composition with no motion.

Picture 4.10

Pict. 4.10

The YouTube video https://youtu.be/sIP6VjcG64U
and the picture below show the impact of the complex magnetic field. The electromagnets are covered by a CD disc during the tests. It blows away the nail indicator from the centre.

There is no voltage within the solenoids: The nail-pointer stands stable within the centre of the disc.	The solenoids are under voltage: 37V The nail-pointer is blown out from the centre and is in chaotic circulation, but strickly around the middle area under the impact of the magnetic blow. There are two moments of the circular movement pictured in two opposite positions.
Picture 4.11/a	Picture 4.11/b,c

Pict. 4.11

The electromagnet-composition and the nail indicator are in similar vertical position on Pictures 4.12 here below, just the centre of the disc is filled with small slivered metal mix. The experiment shows, that the nail is pushed out from the middle, from the area above the internal solenoid, but the metal mix consumes certain intensity capacity and the nail-pointer is blown away with less power:

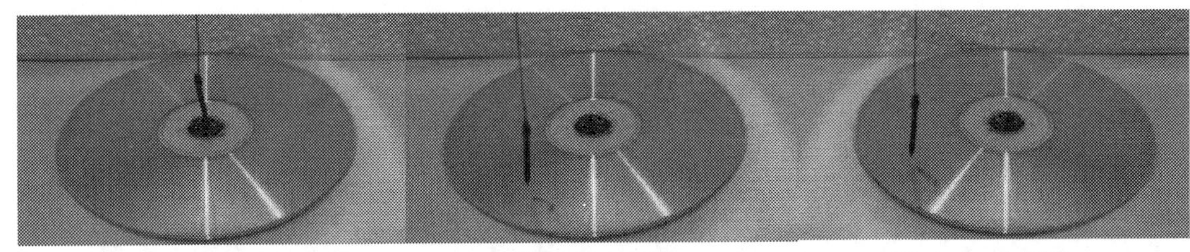

Pic. 4.12 Picture 4.12/a Picture 4.12/b Picture 4.12/c

The experiment with the magnetic drive will be introduced separately in Section 9.

S. 4.1

4.1 Other magnetic impacts

The following experiments were made for demonstrating, that the electromagnetic effect of the anti-electron processes is impacting all elementary processes. The experiments demonstrate it in the case of various elementary processes and subjects. The difference is only in the ways of the reaction, the power and the detection of the impacts.

plastic ring; polyester sheet; fine flour; carbon; concrete and wood.

The dynamism of the magnetic impact was less and the detection difficult in the case of the carbon, concrete and wood targets as presented on Picture 4.16.

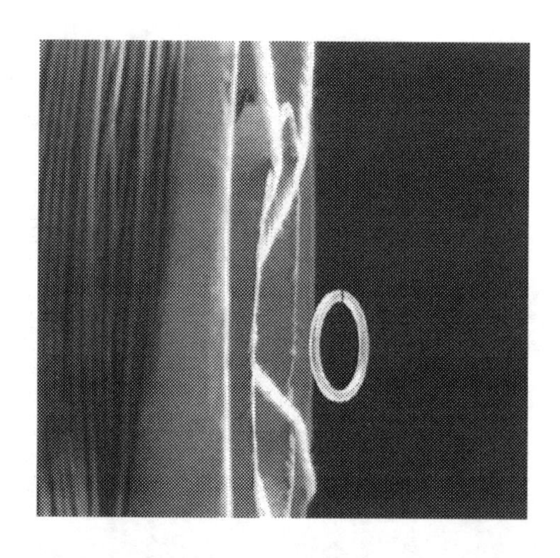

1
The plastic ring is rotated by the magnetic wind

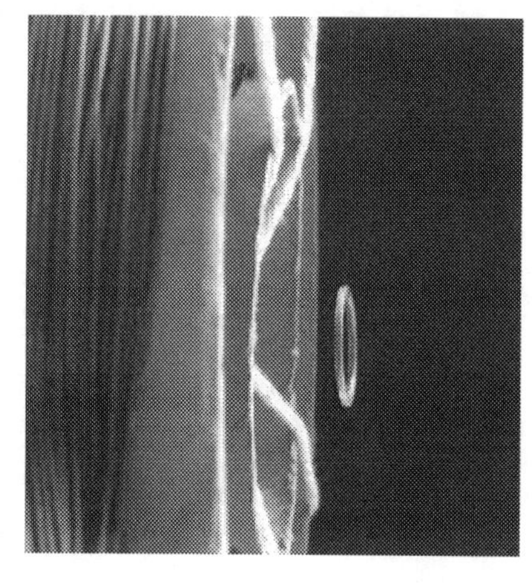

Picture 4.13

Pic. 4.13

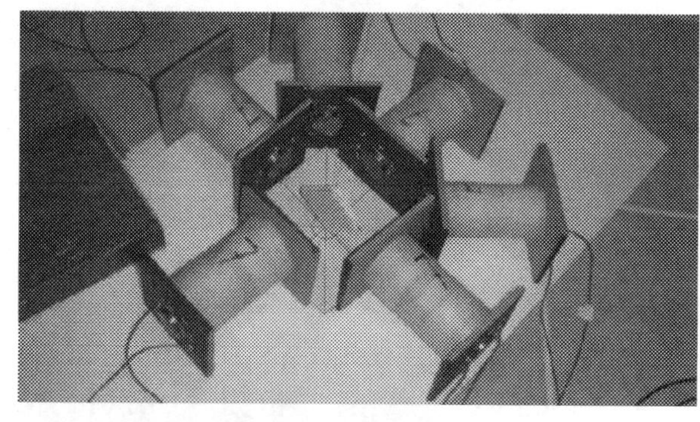

2.
The small *polystyrene sheet* was also rotating and swinging under the impacts of five iron core electromagnets.

Picture 4.14

Pic. 4.14

3.
The electromagnetic impact was also tested spreading *fine flour* from above.
The one on the left is without the impact of the electromagnet.
The same spreading but under the electromagnetic impact is on the right.

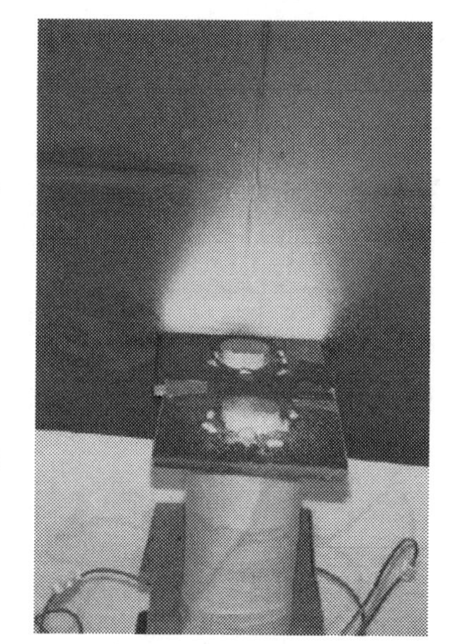

Picture 4.15

Pic. 4.15

carbon

Picture 4.16/a

concrete

Picture 4.16/b

wood

Picture 4.16/c

Pic. 4.16

5
Density, the appearance of the elementary processes in our space-time

The elementary processes are built up from two, opposite to each other processes:

- the sphere symmetrical expanding acceleration = the proton and the electron processes,
- the sphere symmetrical accelerating collapse = the neutron process.

The sphere symmetrical expansion is the source of the internal energy of the elementary process.

The sphere symmetrical collapse is the utilisation of the internal energy of the elementary process. The drive of the collapse is the electron process.

The full cycle includes the processes on the anti-side as well.
The anti-side is the re-generation of the energy during the anti-neutron process. The anti-electron process as the drive of the anti-proton collapse and the utilisation of the re-generated energy in the anti-proton process. Each new elementary cycle starts without the missing quantum impulses of infinite low energy intensity, generating in the direct and the anti-sides within each cycle.

The number of the proton, electron and the neutron processes and also the processes of the anti-sides characterise the intensities of the elementary processes. The Periodic Table is built up on these elementary data.

The integrated results of the intensities of the expansion and the collapse, on both, the direct and the anti-sides are in fact the appearances of the elementary processes in our space-time. The measurement of the appearance is expressed in the certain, unit space value of our space-time: cubic decimetre, dm^3, cubic metre m^3 and others. This parameter of the appearance is the *density* of the elementary process: the (power) impact of the elementary process in our space-time.

D – is the *density* is the physical connection of the two space-times.

The collapses on the two sides are driven by the electron and the anti-electron processes. This way the intensities of these processes are the drives of the appearance.

The pressure of the quantum membrane defines the intensities of the collapses.
The quantum membrane means the acting intensity of the anti-electron processes.
The quantum membrane is the one establishing the intensity of the electron process.
The pressure of the quantum membrane guarantees the identity of the elementary process.

qp – is the pressure of the *quantum membrane*, the key of the appearance (the density) of the elementary process.

$Pa = 1\frac{N}{m^2} = 1\frac{kg \cdot m}{s^2 m^2} = 1\frac{kg \cdot m \cdot m}{s^2 m^2 \cdot m} = 1\frac{J}{m^3}$; $D = \frac{kg}{m^3} = [\rho]$; 5A1

$Pa = 1\frac{N}{m^2} = 1\frac{kg \cdot m}{s^2 m^2} = 1\frac{kg \cdot m \cdot m}{s^2 m^2 \cdot m} = \frac{kg \cdot m^2}{s^2 m^3}$;

This means for the certain elementary process: $qp[Pa]f(\varphi) = D_x \left[\frac{kg}{m^3}\right] * c_x^2 \left[\frac{m^2}{s^2}\right]$; 5A2

$f(\varphi)$ – is the supposed function of more parameters of the certain (x) elementary process, influencing the generation of the pressure value of the quantum membrane.

From the earlier studies of the elementary processes it follows, that

the energy intensity of the proton process (the expansion) is: $e_p = \frac{dmc^2}{dt_p \varepsilon_p}\left(1 - \sqrt{1 - \frac{i^2}{c^2}}\right)$; 5A3

the energy intensity of the neutron process (the collapse) is:

$$e_n = -\frac{dmc^2}{dt_n \varepsilon_n}\sqrt{1 - \frac{(c-i)^2}{c^2}}\left(1 - \sqrt{1 - \frac{i^2}{c^2}}\right);$$ 5A4

ε_p and ε_n are the coefficients of the intensities of the proton and the neutron processes. The relation also characterises the appearance of the elementary processes.

$$\varepsilon_{ex} = \frac{\Delta t_{xn}}{\Delta t_{xp}}\Delta\sqrt{1 - \frac{(c_x - i_x)^2}{c_x^2}}; \quad \text{and} \quad \varepsilon_{xe} = \frac{1}{\varepsilon_{xe-}};$$ 5A5

There is no space-time difference in the electron processes. This is the reason the electron processes communicate without problem: $\Delta t_i = \frac{\Delta t}{\sqrt{1 - \frac{i^2}{c^2}}}$; 5A6

The time count in 5A3 and 5A4 mean the *acting* time of the electron and the anti-electron processes, the drives of the collapses in the elementary process. The entropy products of the drives (the quantum impulses) remain and are parts of the space-time of the electron and anti-electron processes.

Once the quantum pressure, developed by the intensity of the anti-electron process is corresponding to 5A2,

- the case of the relation of $\varepsilon_e > \varepsilon_{e-}$, means $\varepsilon_e > 1$ electron process surplus, like in the cases of the elementary processes *H, He, O, N, Si, S, Ca, C;* 5A7
- the case of the relation of $\varepsilon_e < \varepsilon_{e-}$, means $\varepsilon_{e-} > 1$ anti-electron process surplus, the source of the quantum pressure, like in the case of all other elementary processes. 5A8

The intensity of the anti-electron process defines the ***qp*** and the value of the intensity is $\varepsilon_{xe-} = 1 + \Delta_x$; with the released surplus (the magnetic impact) of the elementary process. 5B1

- The case of the diamond means density of the extra high value and extra stable elementary process with $\varepsilon_{e\ dimond} \leq \varepsilon_{e-\ dimond}$. 5B2

The division of the absolute value in 5A2 by the periodic (***PN***) number of the elementary process gives the unit value of the quantum pressure for the elementary process. Ref. 5A2

The periodic related unit values make the comparison of the different elementary processes possible.

As the diagrams of the chapter prove, the pressure of the quantum membrane is crucial for the conductivity of the elementary processes. As the electricity is the propagation of the *blue shift* quantum impacts of the electron processes (handing them over and over by the elementary processes), the elementary processes with the increased values of the pressure of the quantum membrane have the most efficient transport characteristics.

The Periodic Table is about the appearance of the elementary processes, built up on the existing differences.

All elementary processes conduct the electricity in some extent. The conductivities – the intensities of the communication of the *blue shift* impacts – however are of broad scale. As the assessment of the pressure of the quantum membrane – based on the measured data of the densities – proofs, the higher the pressure value of the quantum membrane is, the best is the conductivity of the elementary process

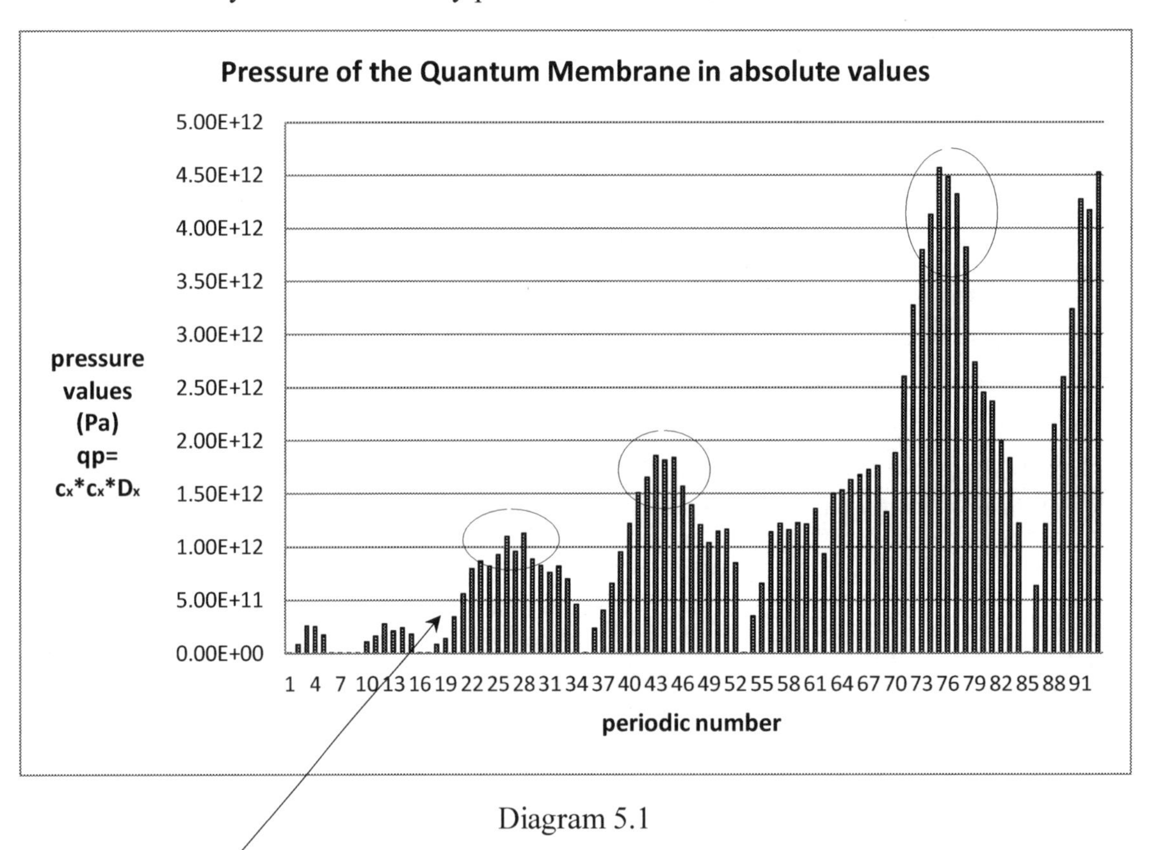

Diag. 5.1

Diagram 5.1

The best conductors in *practical terms* are on the top of groups 7, 8, 9, 10, 11, transition metals. The elementary processes *Co, Cu, Ni, Fe* and *Mn.*

The *Cobalt (Co)27, Cuprum (Cu) 29, Nickel (Ni) 28, Iron (Fe) 26* and *Manganese (Mn) 26* elementary processes have increased values of the pressures of the quantum membranes.

The regions of the *Technetium* (Tc), *Ruthenium* (Ru), *Rhodium* (Rh), *Palladium* (Pl) *Silver* (Ag), with the periodic numbers from 43 to 47; the area between *Tungsten* (W) 73 – *Gold* (Au) 79 all of the *transition metal* group; and around the *Uranium* (94) processes of the *actinide group* have even higher absolute quantum membrane pressures.

The pressure of the quantum membrane, divided by the periodic number gives the *unit values* of the pressure of the quantum membrane.

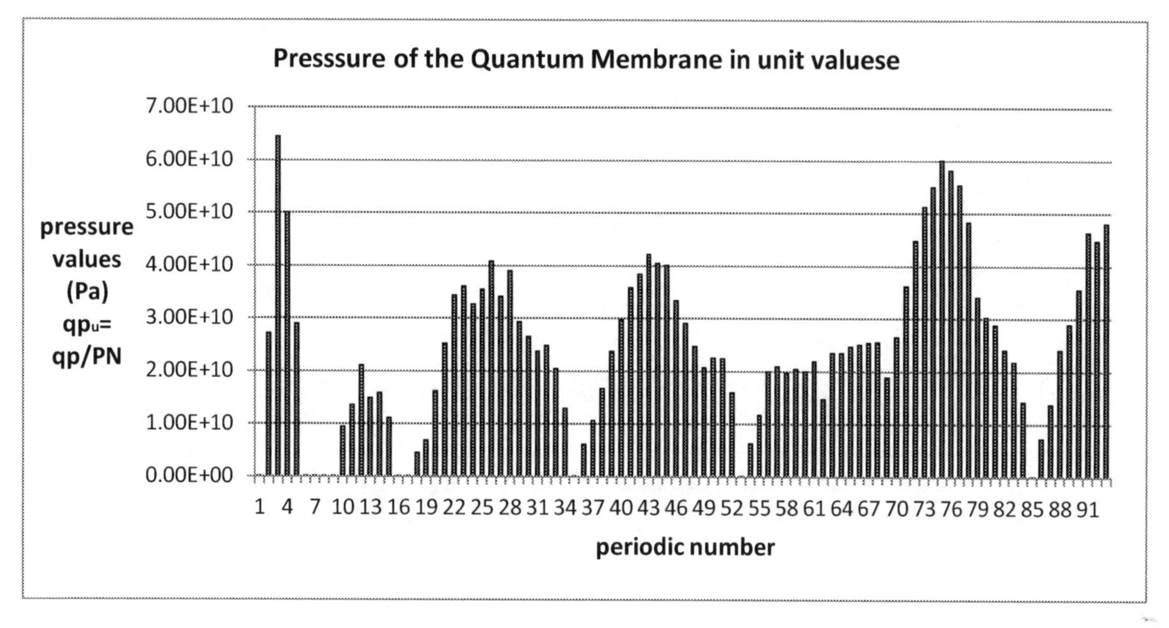

Diagram 5.2

Diag. 5.2

The only surprise in Diagram 5.2 is the high pressure value of the *Beryllium* (PN4) and the *Boron* (PN5) processes. This is the explanation of their positions in the Periodic Table.

The measured density values correlate well with the absolute values of the quantum membrane pressure of Diagram 5.1. The measured density values are presented here below on Diagram 5.3 for the comparison with Diagrams 5.1 and 5.2.

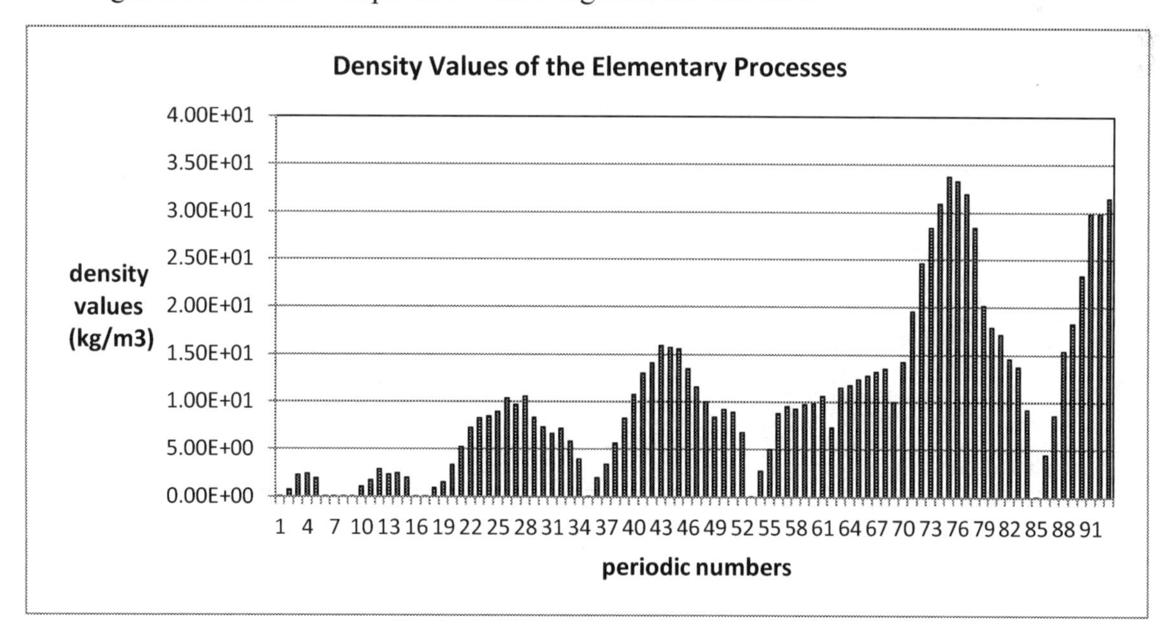

Diagram 5.3

Diag. 5.3

There are significant differences in the values of the pressures of the quantum membranes between the gases and the other aggregate statuses. The comparison, with reference to Diagram 5.1 and Diagram 5.4 below, shows that the difference is in the order of one and two magnitudes. This explains the difference in the aggregate statuses. The unit pressure values for the elementary processes of the gases closely equal:

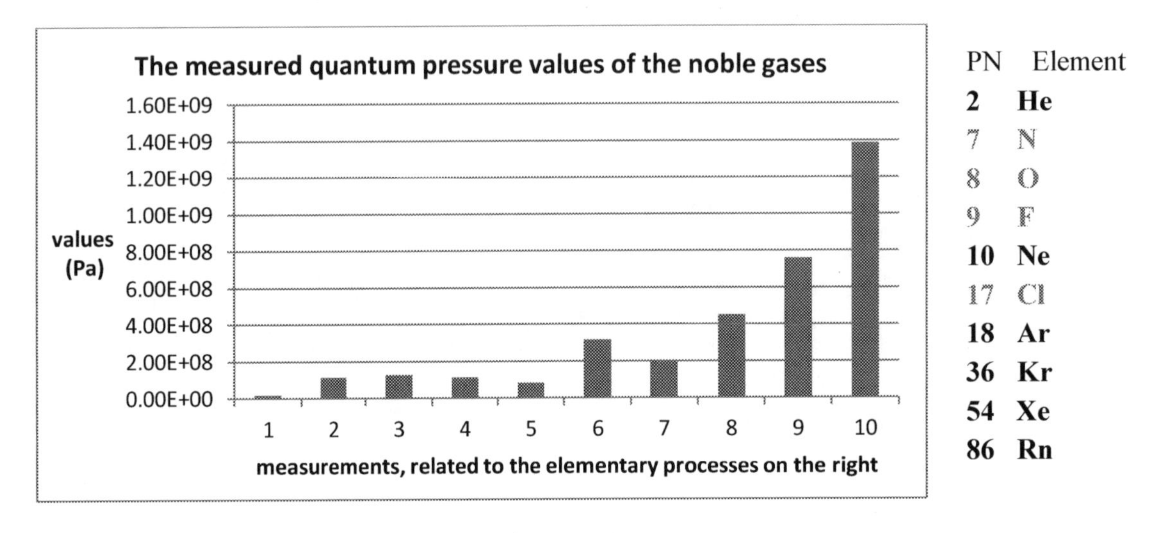

Diag. 5.4

Diagram 5.4

The division of the absolute values of the pressures by the intensity coefficients of the anti-electron processes in Diagram 5.5 demonstrates the stabilities of the structure of all the elementary processes of the Periodic Table.

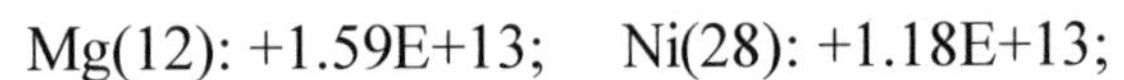

Mg(12): +1.59E+13; Ni(28): +1.18E+13;

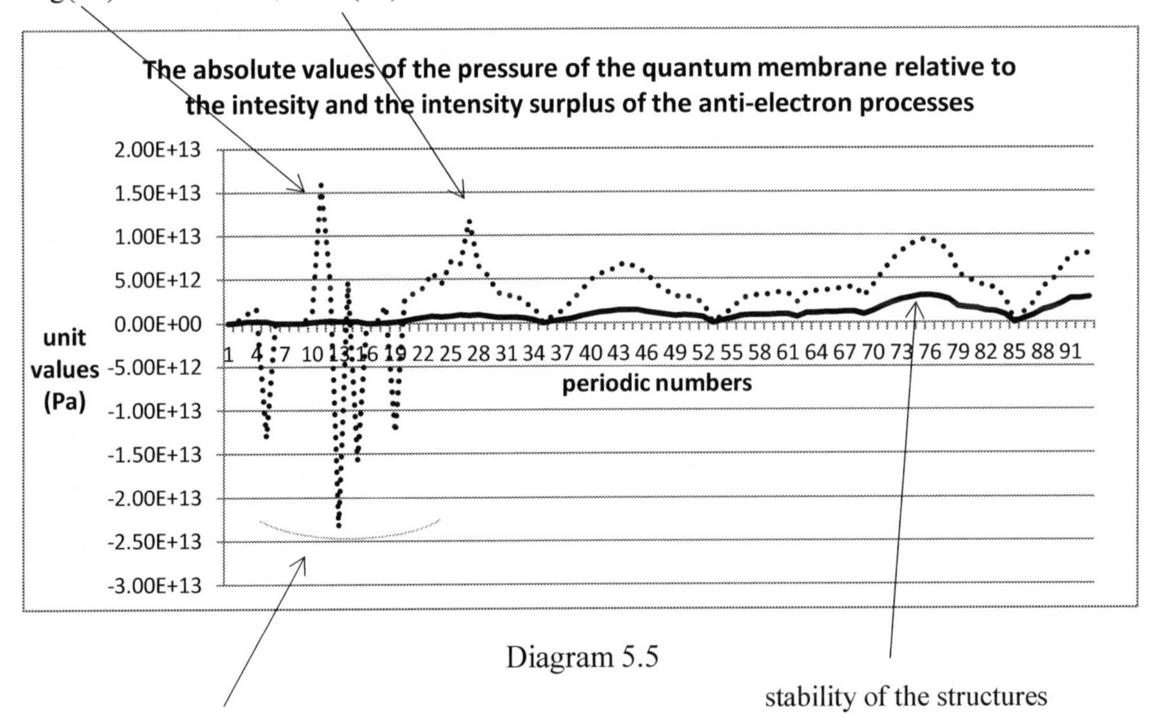

Diag. 5.5

Diagram 5.5

stability of the structures

C(6): -1.32E+13; Si(14): -2.34E+13; S(16): -1.58E+13; Ca(20):-1.26E+13;

The highest surplus of the anti-electron processes belongs to the *Magnesium* and the *Nickel* processes. These are the elementary processes with the highest volume of the *magnetic impacts*.

The increased magnetic impacts of the *Magnesium* – as its name also demonstrates it – and of the *Nickel* processes means stability/protection towards/against the external elementary impacts of the external environment.

The highest volume of the electron process surplus belongs to the most intensive for communication elementary processes.

5.1 The Periodic Table assessed on the unit pressure values of the quantum membrane

S. 5.1

The density values (generated in fact by the quantum membrane) give chance to look at the Periodic Table in different way.

The Diagram 5.6 on the density values, and Diagrams 5.1. and 5.2 on the acting quantum membrane themselves have their dynamic characters and relations.

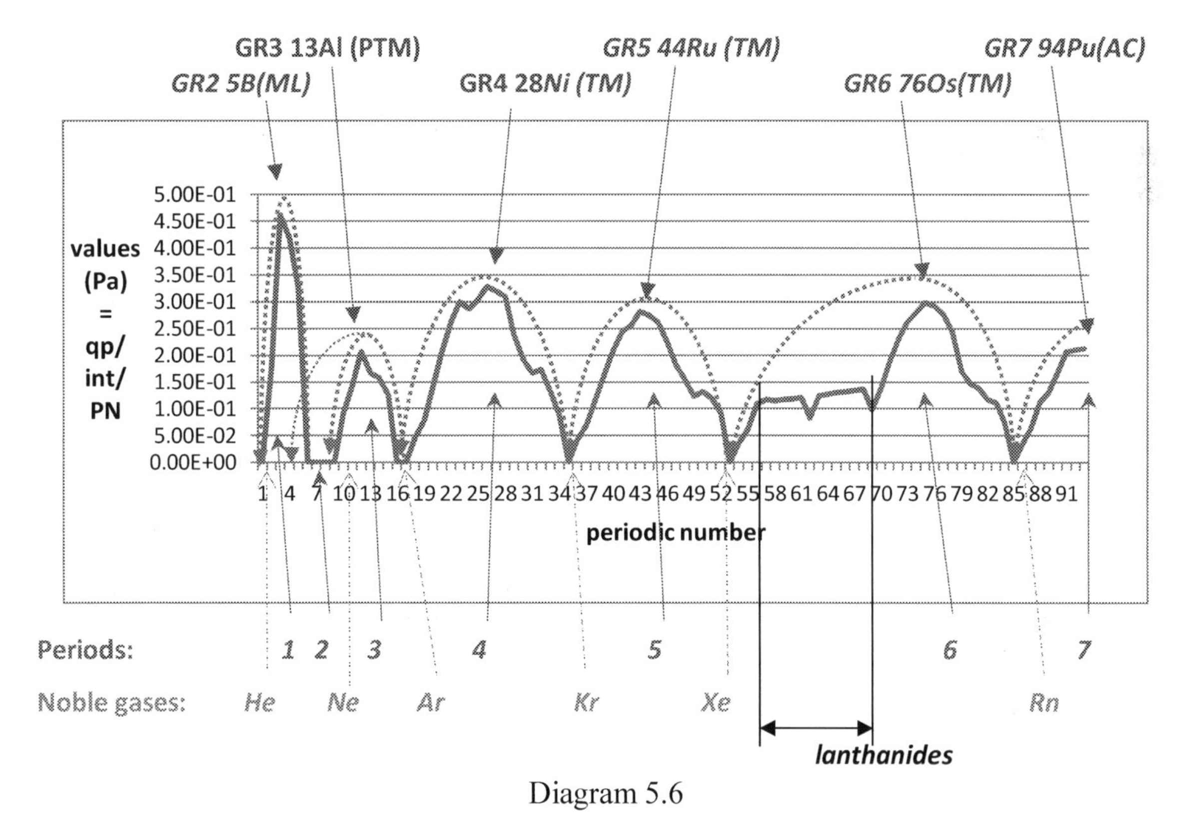

Diagram 5.6

Diag. 5.6

Each period in the Diagram 5.6 has its quasi sinusoid format.

And the sinusoid formats are separated by the elementary processes of the group of the *noble gases.*

The elementary evolution starts at the highest quantum membrane level and expanding step by step. There is a highest quantum membrane value in each period, which divides the period for increasing and decreasing segments.
(ML = metalloid; PTM = post transition metal; TM = transition metal; AC = actinide).

The diagram proves the dominance of the conductive character of the metals.

6
The mechanical impact of the light

The light is the visible appearance of the conflict of the *blue shift* quantum impacts of the electron processes. The conflict also generates heat. The light has its certain intensity and propagates in the space.

The light is the quantum information, generated within the conflict of the elementary processes.

The definition of the light/information/matter might also be given in its reversed format: It is not just the product of the elementary processes, but has its impact on the elementary processes as well. The light signals, with their intensities and frequencies have their own mechanical impacts!

The mechanical impact in its most usual format is force impact. But there is no force impact without the definition of the mass, the appearance of the elementary processes in our space-time. The mechanical effect of the intensities and the frequencies of the light signal can be measured and presented by experiment,

The scheme of the experiment is:

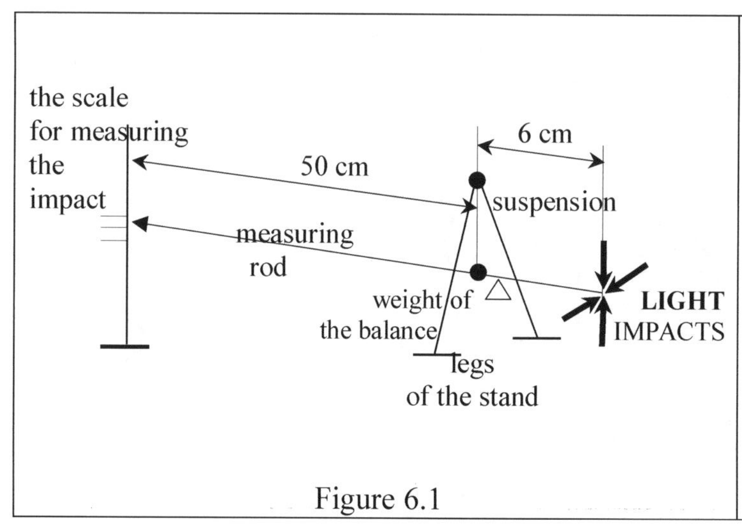

Figure 6.1

The mechanical effect of the light was measured with a simple instrument:
A rod pointer, balanced and suspended is impacted at its shorter end by four symmetrically positioned light signals.

The movement of the other, longer end of the pointer against a scale has proved the impact.

Fig. 6.1

The movement of the pointer of the balanced rod against the scale at one end shows the *force/mass/weight impact* of the light signals on the other end.

Pict. 6.1

Picture 6.1

The light impact, directed on the shorter end of the balanced wire-rod breaks the balance of the rod; the pointer on the other end indicates the shift against the scale.

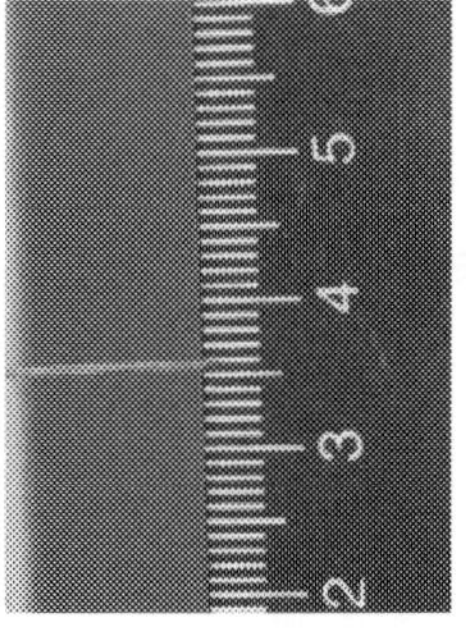

Pict. 6.2

Picture 6.2

The four light signals, positioned symmetrically and perpendicular to each other and focused on the rod are impacting the elementary processes of the shorter end of the rod. The summarised impact has its mechanical effect and sways the other end of the rod, in fact in all directions.

Pict. 6.3

Picture 6.3

The sensor-pointer feels the broken balance and indicates – as above – the impact against the scale. The different lengths of the two arms of the balance exaggerate the movement of the sensor-pointer.

The sway in all directions, value of 0.2-0.3 mm on a 500 mm long arm can easily be detected. The sensor-pointer has its full freedom to move in any direction.

The instrument is placed into a box covered by plastic glass on all sides for preventing it from external impacts.

Pict. 6.4

Picture 6.4

The impact on the shorter end of the rod is clearly mechanical. The mechanical effect is the result of the conflict of the electron processes of the elementary processes of the rod, caused by the quantum impacts of the light signals. The conflict results in the intensity increase of the electron processes and the intensity increase means the increase of the density, the change of the appearance of the elementary processes of the subject in our space-time. The increase of the density results in the disruption of the mass/weight balance of the rod. The *gravitation* is impacting the *mass* and results in the *weight* of subjects.

The increased weight of the sensor-end of the shorter arm, impacted by the light signals means the increase of the *weight*.
The experiment is presented on the YouTube: https://youtu.be/4LY-3PHpF0Q

6.1 The speed of light

S. 6.1

The light is a signal/information of the conflict of the elementary processes. All light signals are generated in elementary conflicts. But not all elementary conflicts are capable and have the intensity, necessary for the generation of the light impact. The light signal is the result of the conflict of the electron process *blue shift* quantum impacts.

The electron processes are the drives of the collapse of the neutron processes. The conflicts result in the increase of the intensity of the quantum drive. As consequence, the collapse of the neutron processes and neutron/anti-neutron inflections become of increased intensities; the generation of the intensity surplus of the anti-electron processes becomes more than the standards. The balancing function of the quantum membrane however restores the intensity of the electron process and the neutron collapse.

The generation of the permanent light signal needs permanent conflict, the permanent *blue shift* quantum impact of the electron processes. The permanent conflict goes with permanent corrections, resulting in the permanent sacrifice of the anti-electron process *blue shift* quantum impacts, the permanent, but increased release of the surplus. The light, sooner or later eats off the impacted elementary processes.

The conflicting electron processes are effecting also the surrounding environment. The consequence of the conflict is the generation of heat and light. The light and the heat signals are transported by the elementary processes or in specific cases by the quantum space. There are however differences in the ways of the transport and the propagation of the light signals, as there are elementary processes with electron process surplus and others with anti-electron process surplus.

The propagation of the light signals in the space with elementary processes means additional conflicts. The conflicts result in the loss of the intensities of the light signals and in the slowdown of the propagation. Therefore the speed values of the propagation of the light signal in the air, water or in other mediums are different than within the vacuum.

The propagation of the light signals within the gaseous and the liquid aggregate statuses is easier than in the elementary processes with solid aggregate statuses. The gaseous and liquid statuses have increased electron process surplus, the solid statuses do not.

Each elementary process has its own electron process intensity value and own space-time. The intensity is established by the anti-electron process quantum membrane.

6A1 The quantum drive of the anti-electron processes for all elementary processes is equal: $\frac{c_x^2}{\varepsilon_{x-}} = const;$

6A2 The quantum drives of the electron processes for all elementary processes are different, as in this case the rule is: $c_x * \varepsilon_x = const$

and while $\frac{1}{\varepsilon_{x-}} = \varepsilon_x$ but $\varepsilon_x \neq const$ - they are for all elementary processes are different,

and $c_x \neq const$ - they are for all elementary processes are also different.

The coefficient of the *intensity* of the electron process (ε_x)

6A3 represents the intensity of the elementary process $\frac{\varepsilon_{px}}{\varepsilon_{nx}} = \varepsilon_x$ and $\frac{1}{\varepsilon_x} = \Delta t_x = (dt_x)$

This means the speeds of the propagation of the light signals depend on the space-time of the elementary process.

This directly means that the duration of the propagation is the function of the intensity of the space-time of the elementary process.

The quantum impulses of the space have no impact on the speed of the propagation of the light. They just convey the quantum information, as the quantum drive for all quantum

6A4 impulses (of the space) are equal and: $\lim \frac{dmc_x^2}{dt_i\varepsilon_x}\left(1 - \sqrt{\frac{(c_x - i_x)^2}{c_x^2}}\right) = 0;$

While the expression above for all quantum impulses is zero limited, the $\frac{c_x^2}{\varepsilon_x}$ component identifies the impulse.

6A5 Expression $\frac{dmc_x^2}{dt_i\varepsilon_x}$ in the formula gives unique frequency (intensity) value: $\frac{kg\frac{m^2}{s^2}}{s} = \frac{W}{s}$

The speed of the propagation of the light in a space without elementary processes *corresponds to the speed value of the space-time of the generation*, to the quantum speed of the elementary process of the generation. The outer space does not add to nor take off the speed of the light in propagation.

The speed value of the propagation of the light signal on the *Earth* surface corresponds to the composition of the atmosphere on the *Earth* surface. The elementary processes of the atmosphere are conflicting with the propagating light signal.

There are two conflicts: the conflict at the point of the generation and the conflict of the propagation. The propagation continues either until

- the signal loses all capacity of the drive, coming from the supply – step by step fading away; or the conflict at the generation of the signal is over – with its immediate stop.

For distinguishing the *intensity* of the elementary conflicts of the generation from the *intensity* of the conflicts of the propagation, the first may be called as *frequency*. The *frequency* of the conflicts of the generation remains without change. The *intensity* of the propagation is changing and depends, with reference to 6A5, on the volume, on the "mass" of the conflicting elementary processes.

Ref. 6A5

Once the energy source of the propagation is off the light disappears!

S. 6.2

6.2 The deflection of the light signal

Electromagnetic quantum impacts are influencing the propagation of the light signal information!
The way of the impact is demonstrated here below on Figure 6.2. The positions mean:

1. the conflict of the generation of the light signal

2. the permanent conflict of the propagation

3. the length of the propagation is the function of the capacity of the drive from the place of the generation

4. *electromagnetic quantum impact* is diverting the direction of the light

Figure 6.2

1: generation of the light;
2: difference in the intensities between the generation and the elementary process of the space is the drive of the propagation;
3: the trajectory of the propagation,
4: the conflict of the light signal and the quantum impact of the electromagnetic waves results in *deflection* = the light is bent.

Fig. 6.2

There is no change in the intensity of the light signal in the space without elementary processes. The quantum impulses are conveying the signal without any change. In the space with elementary processes the propagating signal is losing on the intensity step by step, finally disappears.
In the case the generating conflict is off, the light signal is off wherever is it, whatever is its format, linear or deflected; there is no difference.

Meeting electromagnetic quantum impacts in the space, the trajectory of the light signal becomes deflected.

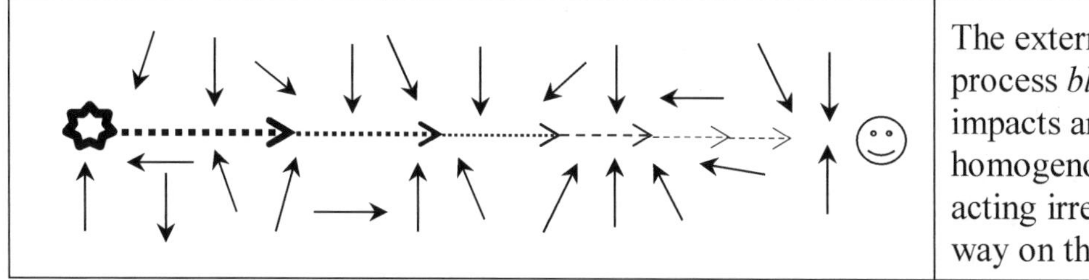

The external electron process *blue shift* impacts are homogenous and are acting irregular, chaotic way on the light signal.

The electron process *blue shift* impacts of the light signal are driven by the light source. There is a certain distance the light signal is capable to pass, even under permanent supply, as it is exposed to the external *blue shift* quantum impact of the elementary processes of the surrounding environment. The permanent conflict results in losses and eats off the energy of the light signal.

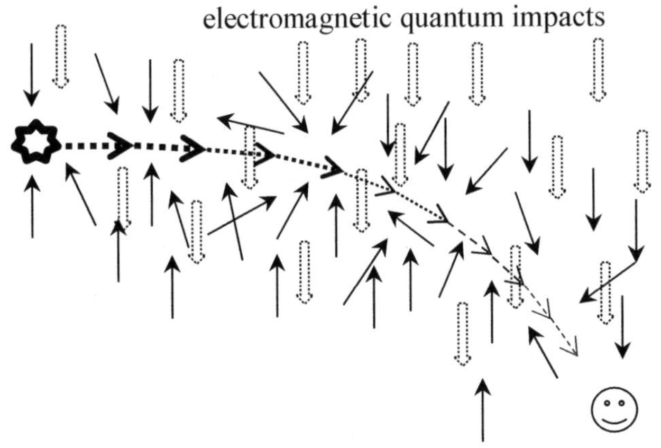

electromagnetic quantum impacts

The light signal is information, conveyed by the elementary processes or by the quantum impulses of the space.

- The magnetic impacts homogenise the chaotic quantum impacts of the electron process of the space with elementary processes and while they also convey the signal, they also mean real conflict within the space with quantum impulses.
- Within the space without elementary processes the magnetic impact and the conflict is direct.

As the results of the conflicts the quantum impacts of the anti-electron process deflect the light signal in both cases!

Figure 6.3

While they pose real resistance and the propagating light signals lose on their intensities, these electron process impacts do not divert the trajectory of the propagation.

But the quantum impacts of the electromagnetic fields are capable for diverting the trajectory of the propagation of the light signal.

Fig. 6.3

The space is built up by the quantum impulses. The intensities of the quantum impulses – as they are the entropy products of the electron and the anti-electron processes – are of infinite low intensity. The quantum impulses do not participate in the elementary communication. The quantum impulses convey the information and signals of any kind, including the electromagnetic signals as well in the space, with elementary processes or without, whatever format or of elementary natures they are.

7
Transmutation

The elementary process is accelerated.

$$\frac{dmc_x^2}{dt_i\varepsilon_x\sqrt{1-\frac{v^2}{c_x^2}}}\left(1-\sqrt{1-\frac{(c_x-i_x)^2}{c_x}}\right);$$ The increased intensity of the electron process is the result of v, the speed of the acceleration from the external source. 7A1

The increase of the intensity of the electron process in 7A1, above, as of ε_{xa} is the consequence of the acceleration: $\varepsilon_y = \varepsilon_x\sqrt{1-\frac{v^2}{c_x^2}}$; 7A2

The increase means: the change of the intensity relation of the proton and the neutron processes and the increase of the surplus of the anti-electron processes. v in the denominator is the actual value of the speed up.

The increased intensity of the electron process results in the intensity increase of the neutron collapse. This generates the increase of the intensity of the anti-electron processes, the generation of the additional surplus of the anti-electron processes.
In order to keep the identity of the elementary processes, the generating surplus and the delta surplus of the anti-electron process, both shall be given off.

If the increased surplus of the anti-electron processes cannot be fully released, the still acting intensity surplus increases the pressure of the quantum membrane and modifies the intensity of the electron process and this way the relation of the intensities of the proton and the neutron processes – the elementary process itself.

The equation in 7A1 might be written in the format of: $n\frac{dmc_x^2}{dt_i\varepsilon_x}\left(1-\sqrt{1-\frac{(c_x-i_x)^2}{c_x}}\right)$; 7A3

or $$\frac{dmc_x^2}{dt_i\varepsilon_x\sqrt{1-\frac{v^2}{c_x^2}}}\left(1-\sqrt{1-\frac{(c_x-i_x)^2}{c_x}}\right)=\frac{dmc_y^2}{dt_i\varepsilon_y}\left(1-\sqrt{1-\frac{(c_x-i_x)^2}{c_x^2}}\right);$$ 7A4

With reference to 7A1, the case might also be understood, that this means the increase of the number of the electron processes, $n=\frac{1}{\sqrt{1-\frac{v}{\varepsilon_x}}}$; 7A5

leading to the *conflict* which is usually escorted by heat generation.
The conflict in this case can only be resolved by the modification of the elementary process, with the increase of the quantum speed ($c_x < c_y$); and with the increase of the intensity of the electron process $(\varepsilon_x > \varepsilon_y)$ = *which is equivalent to transmutation.*

ε_x and the ε_y , are the intensity coefficients (the quotient of the intensities of the proton and the neutron processes).

7A5

The transmutation means in fact the setback in the elementary evolution, resulting in the increase of the intensity of the *proton* process: resulting in a new elementary process with higher speed of the quantum communication and increased intensity of the electron process – with higher periodic number!	$\frac{dmc_y^2}{dt_p}\sqrt{1-\frac{(a_y\Delta t)^2}{c_y^2}}$; a_y is the acceleration of the proton process

Ref. S. 4.1

The electromagnetic field is the impact of the released anti-electron processes. With reference to Section 4.1, all elementary processes can be influenced by the electromagnetic field (exempt the *Hydrogen* process, which does not have anti-electron process) – as the intensities of the anti-electron processes of all elementary processes are equal.

The elementary evolution drives the elementary progress ahead towards the decrease of the speed value of the quantum communication, towards the decrease of the intensity of the electron process. The intensity relation of the proton/neutron processes and the surplus of the anti-electron processes become modified.

The direction of the *transmutation* is opposite to the elementary evolution. The speed of the quantum communication and the intensity of the electron processes are increasing.

The *transmutation* is the change of the elementary structure.

- first of all the increase of the *speed of the quantum communication.*
- as consequence: the *time system*, the frequency of the quantum communication of the elementary process are also changing.

The keys of the transmutation are: the equal values – for all elementary processes – of the products of the square of the speed of the quantum communication and the intensity of the electron process for all elementary processes; and the equal values of the quotient of the square of the quantum speed and the intensity of the anti-electron processes for all elementary processes:

7B1

$$c_x^2 \cdot \varepsilon_x = c_y^2 \cdot \varepsilon_y = \cdots = const \text{ and } \frac{c_x^2}{\varepsilon_{x-}} = \frac{c_y^2}{\varepsilon_{y-}} = \cdots = const$$

Ref. p.50 -52 Fig. 3.1 Ch.3

One of the ways of the transmutation is the increase of the intensity of the electron processes by the impact of an electromagnetic field. With reference to point 50-52 and Figure 3.1 of Chapter 3, the magnetic impact results in the increase of the surplus of the intensity of the anti-electron processes. The pressure of the quantum membrane (the result of the integrated intensity value of the anti-electron processes) is the one controlling the intensity of the electron process.

If the surplus cannot be released, the value of the quantum drive $\frac{c_x^2}{\varepsilon_x}$ and the intensities of the electron processes are increasing. The increase of the intensities of the electron processes, without the correction by the anti-electron process means the change of the relation of the intensities of the neutron and the proton processes – a new elementary process = *transmutation*!

The other way of the transmutation is the increase of the intensities of the electron processes by the mechanical acceleration of the elementary process.

With reference to 7A1, the transmutation needs the increase of the intensity of the electron process, which is equivalent to the increased number of the electron process quantum impacts for the unit period of time (the frequency) with the generation of the temperature to extremely high value:

$$\frac{dmc_x^2}{dt_i\varepsilon_x\sqrt{1-\frac{v^2}{c_x^2}}}\left(1-\sqrt{1-\frac{(c_x-i_x)^2}{c_x^2}}\right); \qquad \text{Ref. 7A1, 7B2}$$

v in the formula means the actual speed value of the acceleration of the elementary process, driven from an external source.

$$\frac{c_x^2}{\varepsilon_x\sqrt{1-\frac{v^2}{c_x^2}}}=\frac{c_y^2}{\varepsilon_y};$$

While the acceleration of the elementary processes increases the frequencies of the quantum impacts for the unit period of the time, it does not change the time systems of the electron processes (t_i). This remains still equal for all the elementary processes. 7B2

The speed of the acceleration of the electron process is $\lim i_x = \lim a_x\Delta t = c_x$;

and $dt_{ix} = \frac{dt_o}{\sqrt{1-\frac{i_x^2}{c_x^2}}}$ 7B3

The third way is the generation of the conflict by the joint quantum impacts of the elementary process in acceleration and the quantum impact of the gravitation:

Therefor the formula of the increase of the intensity of the electron process in 7A1 becomes:

$$\frac{dmc_x^2}{dt_i\varepsilon_x\sqrt{1-\frac{(a\Delta t)^2}{c_x^2}}\sqrt{1-\frac{(gt)^2}{c_{Earth}^2}}}\left(1-\sqrt{1-\frac{(c_x-i_x)^2}{c_x}}\right); \qquad \text{7B4}$$

$(g\Delta t)$ is the quantum impact of gravitation, acting as significant component at the level of the high actual speed (v);
$(a\Delta t)$ is the actual speed of the acceleration;
c_{Earth} is the quantum speed at the *Earth* surface;
ε_x is the intensity of the electron process; and
c_x is the quantum speed of the elementary process, subject to the acceleration.

There is a significant difference between the acceleration of the *Hydrogen* process and the other elementary processes.
As $\lim c_H = 0$ and also $\lim \varepsilon_H = \infty$, there is no way turning the *Hydrogen* process back. 7B5

The *Hydrogen* process remains *uncompleted*, whatever the speed value of the acceleration, the acting electromagnetic impact and the quantum impact of the gravitation are.

While the acceleration of the *Hydrogen* process itself needs a drive of high energy, the developing conflict with the quantum impact of the gravitation results in the additional increase of the temperature – the generation in fact of free energy from the quantum impact of the gravity – the natural source. The higher the speed of the acceleration is, the higher is the temperature of the conflict, the generation of the energy. The subject is discussed in details in Chapter 9.

There given here the examples of the transmutation:

1:

The supposed transmutation of the *Nitrogen* process into *Oxygen* process,

7C1 - the quantum drive of the *Nitrogen* process (PN 7) is 8.808E+10 $\frac{\text{km}^2}{\text{sec}^2}$,

7C2 - of the *Oxygen* process (PN 8) is 8.847E+10 $\frac{\text{km}^2}{\text{sec}^2}$.

The relation of the two is *n=1.0044* for the benefit of the quantum drive of the *Oxygen* process.

7C3 For the transformation of the *Nitrogen* process into *Oxygen* process, the acceleration shall be up to **29,451.0** $\frac{\text{km}}{\text{sec}}$; to be driven from external energy source.

$$\frac{8.808}{\sqrt{1-\frac{u^2}{299670^2}}}=8.847;$$

u = 29,451 km/sec

The value of the speed increase might be less if the acceleration is escorted in parallel by electromagnetic and/or heat impacts.

2:

The supposed transmutation of the *Nickel* process into the *Cuprum* process

7D1 - the quantum drive of the *Nickel* process (PN 28) is: 1.066E+11 $\frac{km^2}{sec^2}$;

7D2 - the quantum drive of the *Cuprum* process (PN 29) is: 1.258E+11 $\frac{km^2}{sec^2}$;

7D3 With reference to 7B2, the relation of the quantum drives gives: $0.847 = \sqrt{1-\frac{u^2}{314007^2}}$

The speeding up of the *Nickel* processes shall reach: **166,749.0** $\frac{km}{sec}$

The conclusion is that the *transmutation* in principle is *possible*!
There might only be chance to make it to happen however by the impact of the magnetic field. There is no other way at the level of our today's technology!

S. 7.1

7.1 The meaning of the heat

The higher the conflict in the elementary communication is, the higher is the intensity increase of the electron processes. The one of the symptoms of the conflict is the generation of heat. It can be formulated in its reversed format, but also correct way as well: the generation of heat means the conflict of the electron processes.

The heat, the quantum impact of the conflict is either propagating away into the free space, or accumulating within the elementary process of the conflicting. The internal intensity increase, (the increased number of the quantum impacts of the electron processes in the unit period of time – the frequency of the conflict), the increasing heat impact may destroy the structure. The higher the frequency of the internal conflict is, the more is the risk of the destruction.

Ch.8

8
The space-time of the pyramids and the red *Copper* process cap on the top

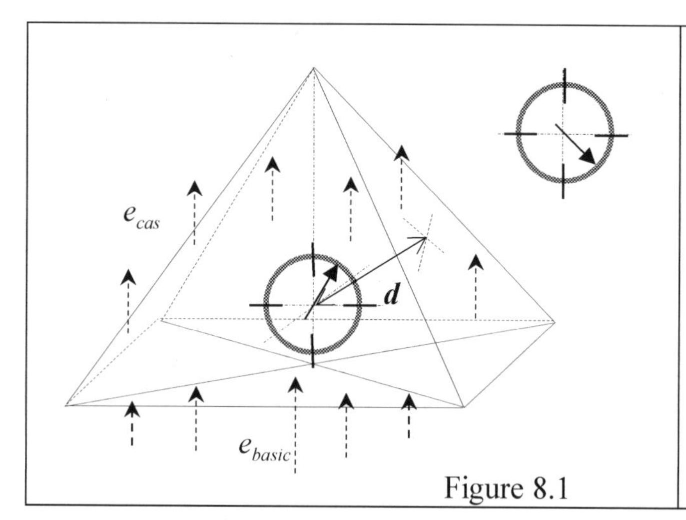

Figure 8.1

The highest intensity inside the pyramid is the area with the highest conflict.

If a clock is placed into this area the clock will be measuring a slower time flow!

The mechanism of the clock is without any change, the reason of the slowdown is the different space-times of the internals of the pyramid and the external environment.

Fig. 8.1

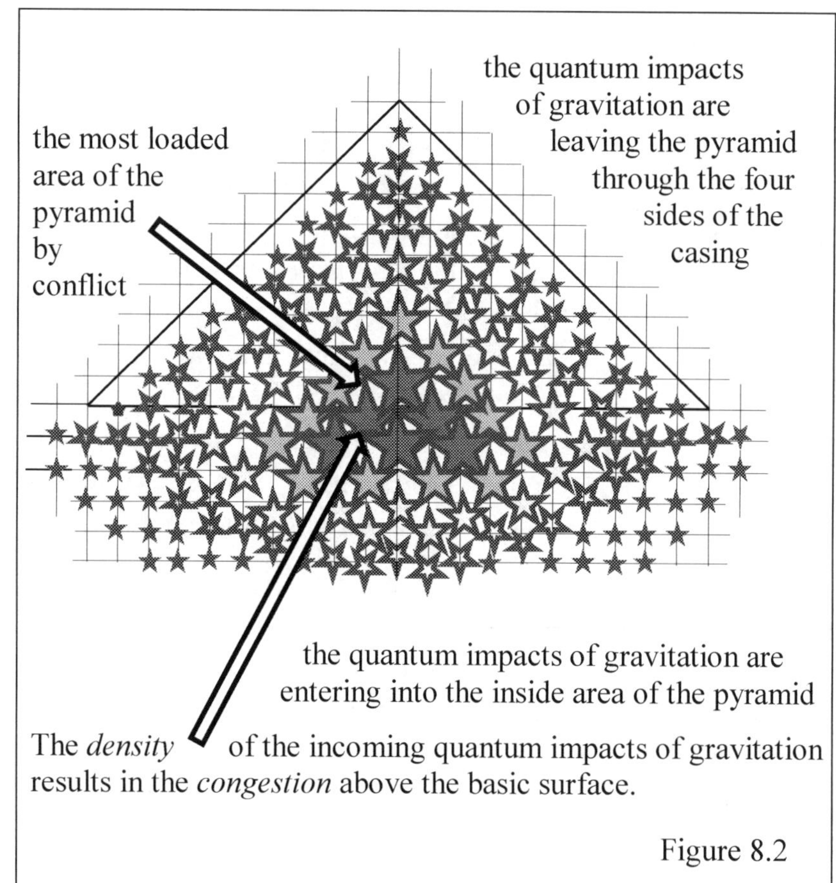

Figure 8.2

The reason of the slowdown is the *increased intensity* of the elementary communication within the pyramid, caused by the conflict of the incoming anti-electron process quantum impact of gravitation.

The *density* of the incoming – through the basic surface – quantum impacts is higher than the density of the released impacts trough the surfaces of the casing.

The conflict is proportional to the relation of the *surfaces* of the four sides and the base.

Fig. 8.2

Any *free* connection of the internals of the pyramid with the external space resolves the time difference – the *space-times* of the internals and the externals communicate.

The case is similar to the example with two platforms in motion: the first is in motion, driven from an external source in certain direction with speed v_e and the other, carrying the clock, moves with speed v_{cl} on the platform of the first in the opposite direction, driven by the energy of the clock itself.

Fig. 8.3

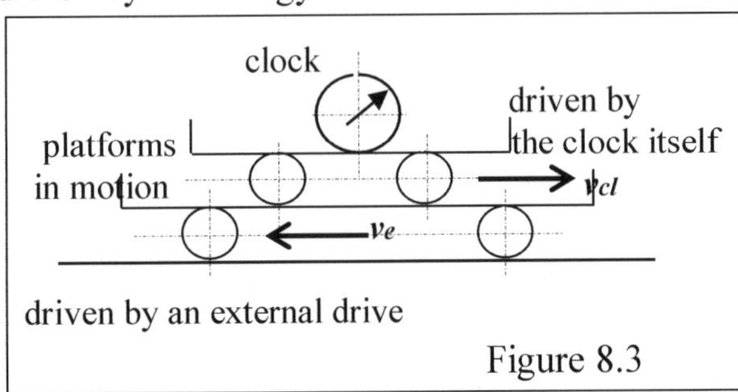

Figure 8.3

If $-v_{cl} = v_e$ there is no time benefit. The intensity increase, result of the external drive becomes lost by the internal intensity demand of the motion of the clock in the opposite direction.

8A1 The intensity benefit of the clock, as result of the motion, driven by an external drive is:

$$\Delta e_+ = \frac{dmc_E^2}{dt\sqrt{1-\frac{v_{ex}^2}{c_E^2}}}\left(1-\sqrt{1-\frac{v_{ex}^2}{c_E^2}}\right) = \frac{dmc_E^2}{dt}\left(\frac{1}{\sqrt{1-\frac{v_{ex}^2}{c_E^2}}}-1\right);$$

8A2 The loss, result of the motion of the clock, driven by its internal energy:

$$\Delta e_- = \frac{dmc_E^2}{dt\sqrt{1-\frac{v_{ex}^2}{c_E^2}}}\left(1-\sqrt{1-\frac{v_{cl}^2}{c_E^2}}\right) = \frac{dmc_E^2}{dt}\left(\frac{1}{\sqrt{1-\frac{v_{ex}^2}{c_E^2}}}-\frac{\sqrt{1-\frac{v_{cl}^2}{c_E^2}}}{\sqrt{1-\frac{v_{ex}^2}{c_E^2}}}\right);$$

if $|v_{cl}| = |v_{ex}|$ the intensity is without change, there is no modification of the time flow;

if $|v_{cl}| > |v_{ex}|$ resulting in the decrease of the original intensity at rest: the measured time is speeding up relative to the time count of the external space (while the rate of the clock remains unchanged);

if $|v_{cl}| < |v_e|$ the result is the increase of the original intensity at rest: the measured time is slowing sown.

With reference to the *pyramid* on Figures 8.1 and 8.2, the explanation for the pyramid with the quantum impact of gravitation is similar. The intensity increase within the pyramid results in the difference in the time count within and outside of the pyramid.

If the internal space of the pyramid is connected with the external environment outside, the case will be similar to 8A2, with the partial or full loss of the gained benefit. The magnitude of the loss depends on size of the window to the external environment. The direct contact with the environment does not cancel the impact, just limits the benefit. There are pyramid impacts acting even a simple house with a roof similar to the pyramid format.

There is always communication between the internal space and the external environment of the pyramid. The reason is, that whatever high the density of the mineral structure of the pyramid is, the quantum communication cannot be completely stopped. The loss can be minimised by the increase of the elementary balance of the minerals, but there is no perfect balance: $\lim E_{comm} \neq 0$. The perfect balance would stop the quantum communication!

Ref. Fig. 8.1

With reference to Figure 8.1, the larger the size of the pyramid is, the more massive is the "protection" of the internal space-time from the surrounding external space-time. Meaning: at equal proportions of the pyramids the longer the distance from the centre of the intensity of the pyramid to the casing is, the less is the loss.

Ref. Fig. 8.3

With reference to the Figure 8.3, the specific similarity between the two cases is that the slowdowns in both cases are results of external impacts. The increase of the intensity within the pyramid is caused by its internal conflict, but the conflict itself is the result of the quantum impact of the gravitation, an external source. In the case of the motion it is the result of the speeding up, driven from external source.

As $c_{py} > c_E$, the speed of the quantum communication within the pyramid is higher than that on the surface of the *Earth*

$$\Delta e_{py} = \frac{dm}{dt}\left(1 - \frac{c_{py}^2}{\sqrt{1 - \frac{v^2}{c_{py}^2}}}\right); \qquad \text{8A3}$$

This requires the best as possible separation from the external environment; and the best as possible relation of the surfaces of the casing and the base.

The more robust the construction is, the less is the relative loss, the higher is the time difference.

The intensity of the quantum impact, in fact the load from the *Earth* is:

$$e_g = \frac{dmc_E^2}{dt_i \varepsilon_{E-}}\left(1 - \sqrt{1 - \frac{(c_E - i_e)^2}{c_e^2}}\right); \qquad \text{8A4}$$

The acting intensity through the surfaces of the pyramid is:

$$e_{py} = \frac{S_{casing}}{S_{base}} \cdot \frac{dmc_E^2}{dt_i \varepsilon_{E-}}\left(1 - \sqrt{1 - \frac{(c_E - i_E)^2}{c_E^2}}\right); \text{ and} \qquad \text{8A5}$$

the relation of the quantum drives of the quantum communication and the gravitation is:

$$\frac{c_E^2}{\varepsilon_{E-}} = \frac{S_{base}}{S_{casing}} \frac{c_{py}^2}{\varepsilon_{py-}}; \qquad \text{8A6}$$

The relation of the *quantum drives* in optimal case gives the *Golden Ratio*:

$$\varphi = \frac{IQ_{py-}}{IQ_{E-}} = \frac{S_c}{S_b}; \qquad \text{8A7}$$

In normal circumstances the quantum drives of the anti-processes are equal (as for gravitation and all elementary anti-processes):

$$\frac{c_x^2}{\varepsilon_x} = \frac{c_y^2}{\varepsilon_y}; \qquad \text{8A8}$$

with reference to 8A7, the surplus of the quantum impact of gravitation, generating the conflict, shall be given off: $IQ_{py-} > IQ_{E-}$ since $S_c > S_b$ 8A9

This means increased quantum speed within the pyramid:

$$c_{py} = c_E\sqrt{\frac{S_c}{S_b} \cdot \frac{\varepsilon_{py-}}{\varepsilon_{E-}}}; \qquad \text{8A10}$$

There are two points here to be noted:

1. The pressure of the quantum membrane, the quantum drive and the quantum speed of the communication are of increased values within the pyramid;
2. The value of the acting quantum drive within the pyramid depends on the rate of the surfaces of the casing and the base.

S. 8.1

8.1
The conflict

8B1 8B2 8B3

The relation of the surfaces of the *casings* and *the cross sections* of classical *Giza* pyramid is equal to:	$\frac{S_{cas}}{S_{cr.sec}} = \varphi = 1.6189;$	$\frac{a}{h} = 1.57$
The relation of the *half of the perimeter* of *the cross sections* and the heights is equal to π:	$2\frac{a}{r} = \pi;$ and $2\frac{a}{\pi} = h;$	
This relation of the surfaces above are slightly differ as the real configuration of the casing is a step by step format:		$\varphi = 1.6366;$

The cross sections in the vertical direction of the small test pyramid are changing. The area of the cross sections is the function of the heights. The relation was measured on a small test pyramid size of $h = 500\ mm$.

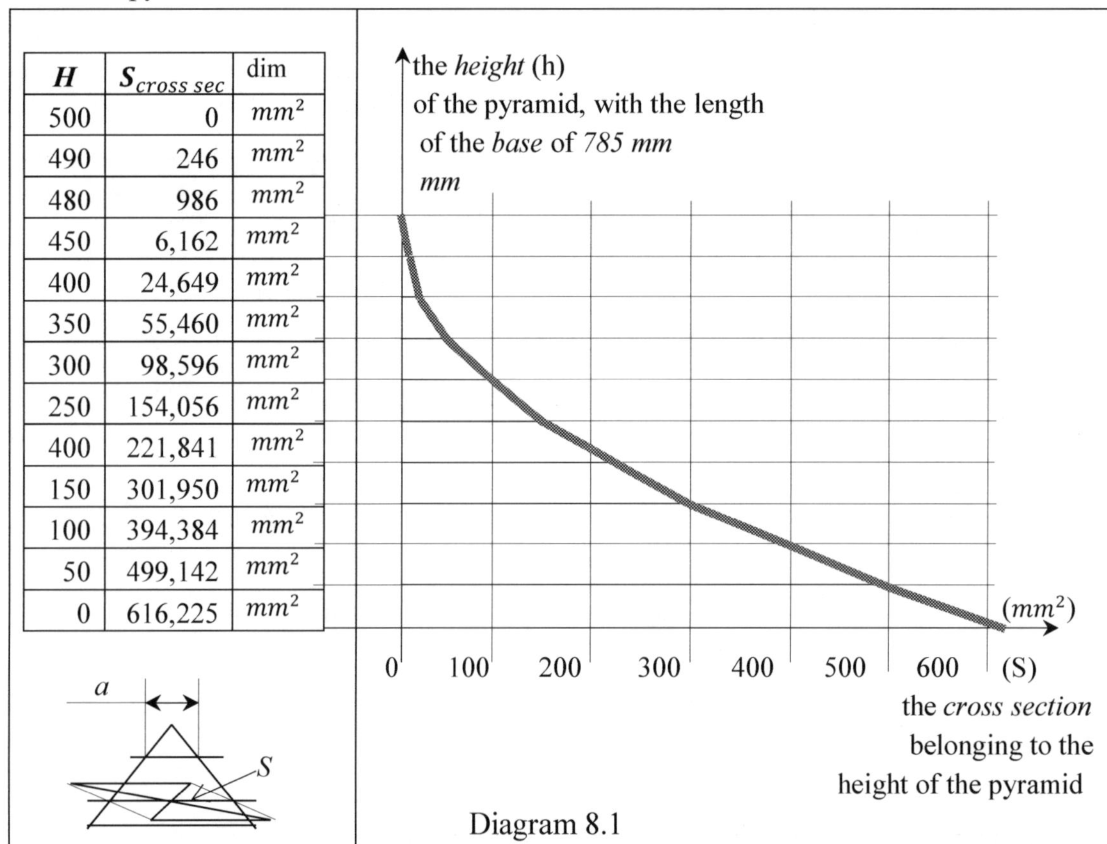

H	$S_{cross\ sec}$	dim
500	0	mm^2
490	246	mm^2
480	986	mm^2
450	6,162	mm^2
400	24,649	mm^2
350	55,460	mm^2
300	98,596	mm^2
250	154,056	mm^2
400	221,841	mm^2
150	301,950	mm^2
100	394,384	mm^2
50	499,142	mm^2
0	616,225	mm^2

Diag. 8.1

Diagram 8.1

The reason of the internal conflict is the difference between the surfaces of the *cross sections* and the related *casings*.

The values of the cross sections and the casings in line with the formula in 8B1 are:

Ref. 8B1

$S_{cr.sec}$	0	246	986	6162	24649	55460	98596	mm^2
S_{cas}	0	399	1564	9976	39904	89785	159617	mm^2

$S_{cr.sec}$	154056	221841	301950	394384	491142	616225	mm^2
S_{cas}	249402	359138	488827	638468	808061	997607	mm^2

Table 8.1

Tab. 8.1

And the differences are:

h	0	100	200	300	400	490	500	mm^2
$S_{cas} - S_{cr.s}$	381,382	244,084	137,297	61,021	15,255	153	0	mm^2

Table 8.2

Tab. 8.2

The values of the cross sections for the *Giza* pyramid are:

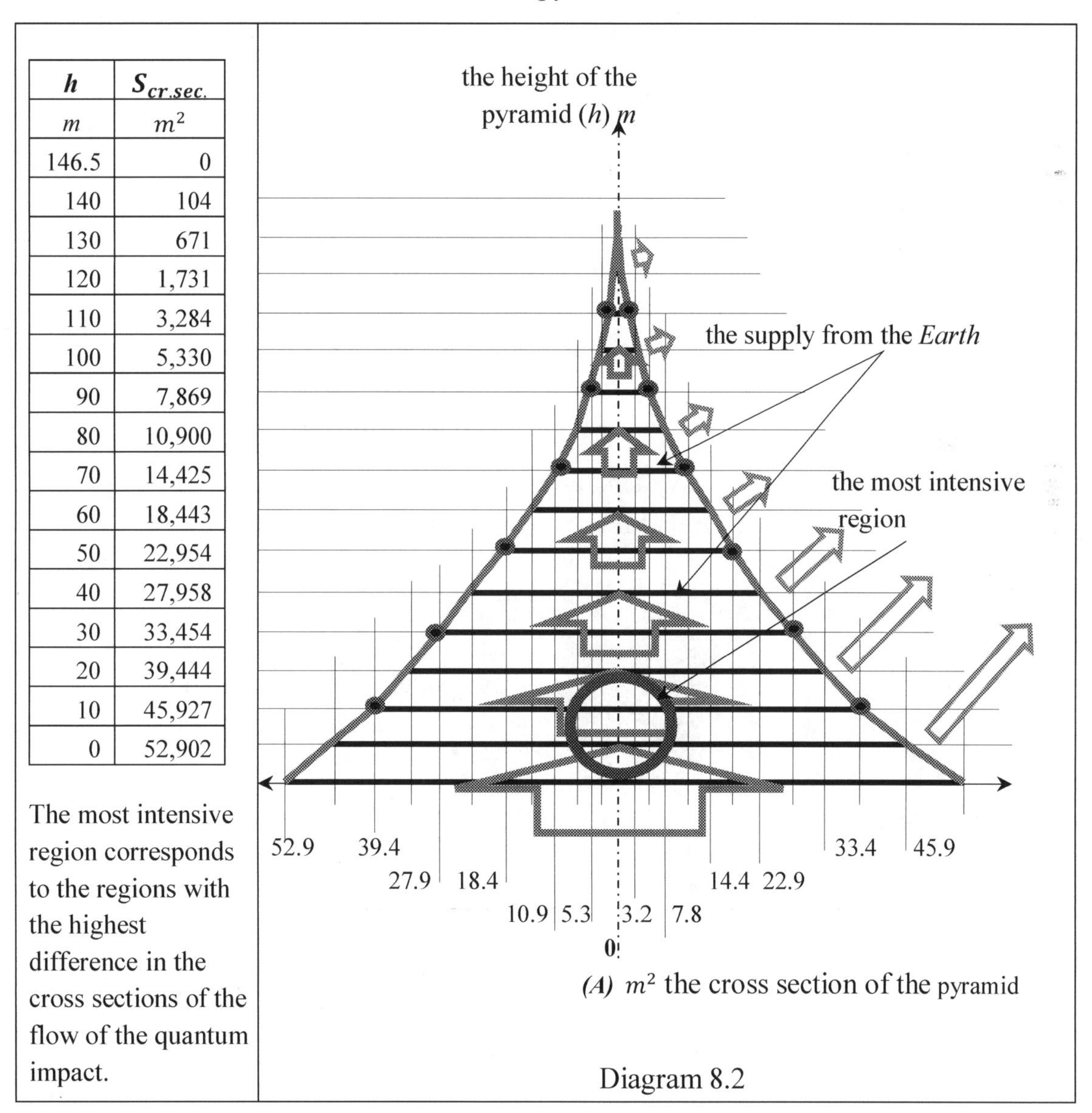

h	$S_{cr.sec.}$
m	m^2
146.5	0
140	104
130	671
120	1,731
110	3,284
100	5,330
90	7,869
80	10,900
70	14,425
60	18,443
50	22,954
40	27,958
30	33,454
20	39,444
10	45,927
0	52,902

The most intensive region corresponds to the regions with the highest difference in the cross sections of the flow of the quantum impact.

Diagram 8.2

Diag. 8.2

The differences, function of the height values well demonstrate the difference between the quantum impacts of the load and the release, as in Table 8.3 on the next page.

$h(m)$	0	20	40	60	80	100	120	146.5	m
$S_{cas} - S_{cr.sec}$	33,696	25,124	17,807	11,747	6,943	3,395	1,103	0	m^2
$S_{case(n)} - S_{cas(n+1)}$	0	8,572	7,316	6,060	4,804	3,548	2,292	1,103	m^2
$S_{cr.sec(n)}$	52,902	39,444	27,958	18,443	10,900	5,330	1,731	0	m^2
S_{supply}	86,598	64,568	45,765	30,190	17,843	8,725	2,834	0	m^2
Congestion		22,031	18,803	15,575	12,347	9,119	5,891		

Tab. 8.3

Table 8.3

S_{supply} $S_{supply} = S_{cr.sec} + S_{cas}$	characterises the *supply* of the quantum impact of gravitation through the cross sections;
$S_{case} - S_{cr.s}$	the surplus of the *flow* through the casing;
$S_{cr.sec(n)}$	the *flow* relating to the cross section;
$S_{case(n)} - S_{case(n+1)}$	the *reduction* of the flow through the casing of two consecutive sections.

Ref. Fig. 8.2

Figure 8.2 demonstrates well the load of the quantum impact of the pyramid. The density of the supply within the central areas has an increased character, consequence of the increased distance to the casing.

The increased density means the congestion of the impacts, in fact the intensity increase, the conflicting quantum impacts of gravitation within the pyramid. The intensity increase also means the slowdown of the time count. The increasing intensity modifies the *space-time* within the pyramid. The most intensive regions are close to the basic surface of the pyramid.

The test pyramids, covered by snow demonstrate well the internal conflict of the quantum impact of the gravitation. The conflict has its typical tent format, resulted with reference to Figure 8.2, Diagrams 8.1 and 8.2, by the internal potential of the pyramids.

Pict. 8.1

Picture 8.1/a and /b

With reference to: https://www.youtube.com/watch?v=T7jOK7yqmA4

This permanent conflict keeps the internals of the pyramid at higher temperature than the environment and the surface of the *Earth*.

This phenomenon can only be explained if the approach is process based, with elementary processes, events, intensities, time, space and others.

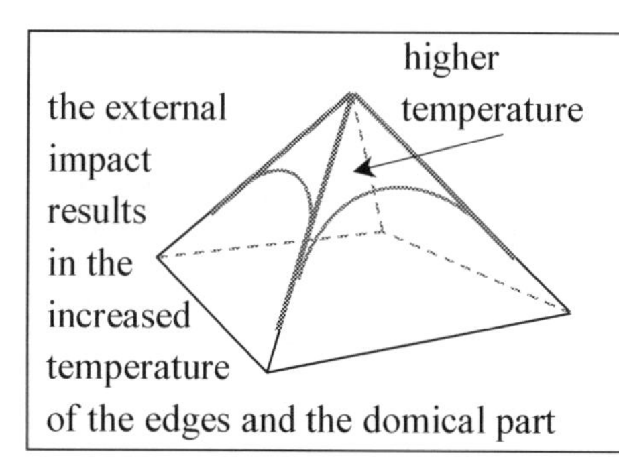

The same phenomenon was measured on the *Giza* pyramid, just with the opposite direction.
The edges and the domical part of the pyramid are far from, and more or less out of the control of the internal conflict. They are therefore exposed here to the external impact with high temperature.
Figure 8.4

Fig. 8.4

With reference to the test pyramid on Picture 8.1/a and /b, the highest temperature within the pyramid is measured at the height of one third from the base. But the internal temperature at the height one third from the top is still higher than the environment. There is no difference is the weather hot or cold. Just the temperatures differ, obviously in the favour of the summer time. The upper part for certain extent and the slim edges are more or less out of the internal control, therefore these are more exposed to the environmental impact, either it is hot around or cold, no difference.

On Pictures 8.1/a and /b: the external temperature is 3.5°C, the surface is 5.3°C, the soil below the pyramid is 2.5°C. The temperature within the pyramid is 12°C at the one third from the base; 9°C from the top. The SNOW covers the upper part – in tent format and the edges on the two third, which are more exposed to the external impact. (The small snow heap on the front side covers the thermometers.)

The value of the electricity generation within the pyramid in winter time is around: $300\ \mu A$. In summertime with the external temperatures around 25°C and with internal temperatures $35-45$°C the value of the generating and released current is of $400-700\ \mu A$.

8.2
Why the granite is the best material for the pyramid?

S. 8.2

The elementary processes of the granite are close to the equilibrium state: the intensities of the proton and neutron processes are quasi equal. The surplus of the anti-electron processes is close to zero.

$$\frac{dmc_{gr}^2}{dt_p\varepsilon_p}\left(1-\sqrt{1-\frac{v^2}{c_{gr}^2}}\right) \cong \frac{dmc_{gr}^2}{dt_n\varepsilon_n}\sqrt{1-\frac{\left(c_{gr}-i_{gr}\right)^2}{c_{gr}^2}}\left(1-\sqrt{1-\frac{v^2}{c_{gr}^2}}\right); \qquad \text{8C1}$$

Therefore the generation of the quantum impact within the granite is: $\lim q_{gr}=0$ 8C2

$$\text{and} \quad \varepsilon_e \cong \frac{\varepsilon_p}{\varepsilon_n}\sqrt{1-\frac{\left(c_{gr}-i_{gr}\right)^2}{c_{gr}^2}} \cong 1\,; \qquad \text{8C3}$$

The elementary composition of the granite in principle does not communicate with the environment. For keeping the continuity of the direct quantum impact of the gravity however the granite has to communicate. The portion, leaving through the casing shall be compensated from below, from the quantum impact of the gravity.

The source of the energy potential of the pyramid is the quantum impact of gravitation. The pyramid, lifted up from the *Earth* surface shows no potential. The more the surface of the *Earth* covered by pyramid is, the more is the generating potential within the pyramid.

Pic. 8.3

Picture 8.3

The quantum impact from the casing shall harmonise with the intensity of the environment. And the load of the quantum impact through the base shall match the demand.

The higher the conflict inside the pyramid is, the higher is the increase of the temperature inside, the higher is the radiated quantum impact through the surface of the casing into the environment. The geometry establishes the relation of the incoming and the leaving quantum impacts.

The relation of the numbers (the intensities) and the densities of the incoming quantum impacts of gravitation and the quantum impact, leaving the pyramid through the casing depend on the proportion of the surfaces of the base and the casing. The experience proves that the potential difference of the pyramid – similar in its size proportions to the *Great Pyramid of Giza* - is around *1V*. The current leaving through the top of the pyramid depends on the size of the pyramid.

The value of the current from the pyramid depends on the potential difference between the top of the pyramid and the other end. The higher the potential difference is, the higher is the value of the current. The measured intensity of the electricity flow from a small pyramid, height of 500 mm is *0.4 mA*; from an even smaller, height of 150 mm, is *0.25 mA*. If it is supposed that the potential difference of a *Giza* type pyramid between the top of the pyramid and the *surface of the Earth* is also around *1V*, the intensity of the electricity flow from the *Giza Pyramid,* 25 million times higher in its size than that of the pyramid with height of 500 mm and might be around 10-15 kA at *1V* potential. The *1V* is far enough to drive the electron process *blue shift* impact through the quantum space (and the atmosphere). The amperage value is high enough for transferring it on higher voltage. The higher the mass of the pyramid is, the higher is the value of the current.

S. 8.3

8.3
The top with the *Copper* process cap and the communication of the pyramids
The healing impact of the pyramids

The experiments prove that there is no need for wires to be built into the pyramid for communication. The test pyramid, with a *Copper* process cap on the top communicates the same way.

There is a test pyramid with a red *Copper* process cap on the top, size of $a = 80\ mm$ and $h = 51\ mm$ on the next picture. The size of the pyramid itself is $a = 880\ mm$ and $h = 560\ mm$. There are no wires built into the pyramid. The *Copper* cap was connected with the surface of the *Earth* and the energy potential of the pyramid was measured.

Picture 8.4/a,b

Pic. 8.4

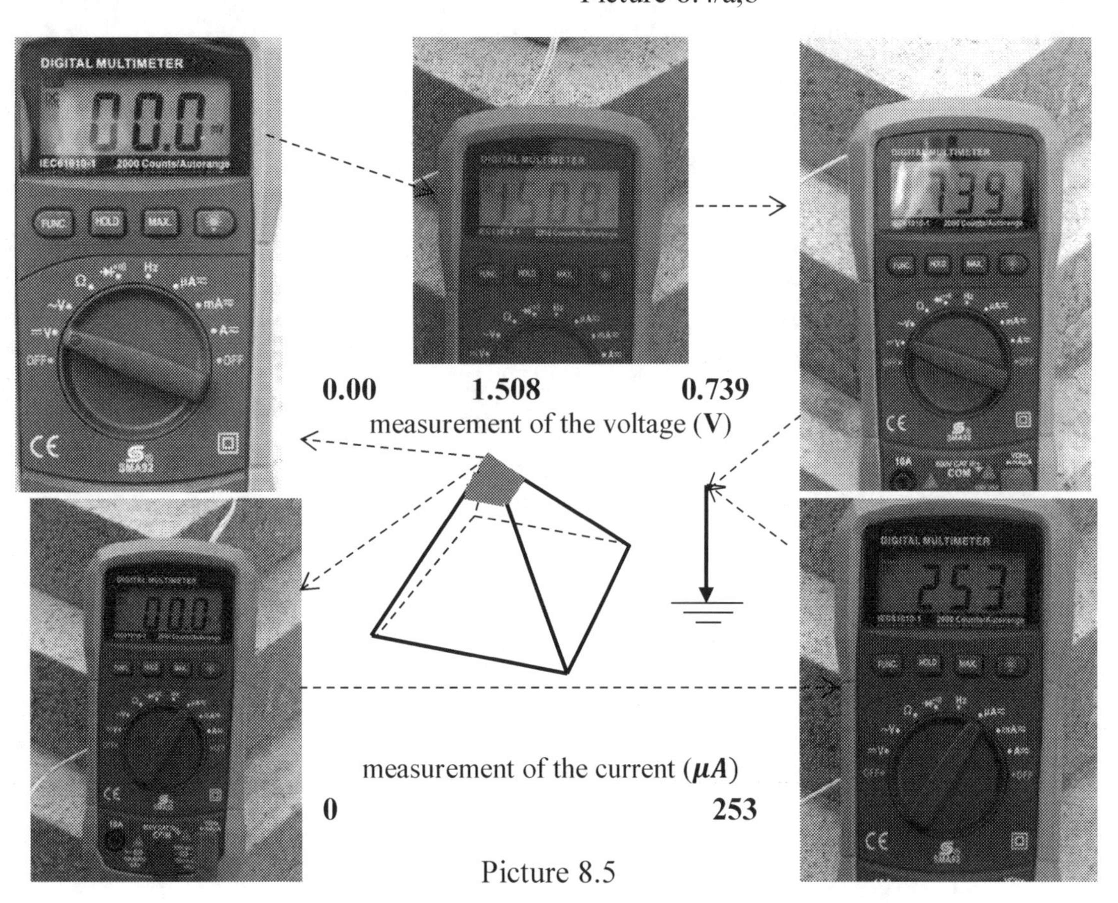

Picture 8.5

Pic. 8.5

The values of the voltage and the amperage were measured as illustrated above.
At the moment of the connection the measured voltage was *1.508 V*, in one second it stabilised on *0.739 V*.
The stabilised value of the current leaving the pyramid was 253 μA.

Picture 8.6 shows the scheme of the measurement: the elementary processes of the *Copper* cap are connected with the surface of the *Earth*.

The permanent quantum impact of the gravitation results in the growth of the pressure of the quantum membrane of the elementary processes of the pyramid.

Pic. 8.6

Picture 8.6

- The increasing pressure of the quantum membrane initiates the intensity growth of the electron process of the pyramid.
- But the elementary processes of the pyramid are close to the balance of the intensities of the proton and the neutron processes. There is in fact no way to increase the intensity of the collapse of the neutron processes.
- The elementary processes of the pyramid are of increased density value, be it granite, or concrete with the mixture of the elementary processes of specific minerals.
- There is a conflict developing within the pyramid under the quantum impact of the *Earth*, with the increase of the internal temperature.
- There is no other way for resolving the conflict, just the release of the surplus of the anti-electron processes through the casing.
- The release of the anti-electron processes means the impact on the external magnetic field.
- In the meantime the impact of the increasing pressure from below is reaching the red *Copper* (or the *Gold*) processes on the top as well, both – with reference to Chapter 5 – of high values of density and increased values of conductivity.
- The impact is similar: the growth of the pressure of the quantum membrane initiates the growth of the intensity of the electron process drive for the increase of the intensity of the collapse of the neutron, which means the increase of the density of the elementary process.
- All elementary processes have their own density values. In the case of the *Cuprum* process the room for reaching the critical point of the conflict is more for than for the granite.
- Having the conflict, the elementary processes on the top release the surplus of the anti-electron processes and contribute to the impact on the magnetic field of the pyramids.
- The conductivity of the elementary processes of the red *Copper* is of increased value.
- The connection of the *Copper* process with the surface of the *Earth* results in the release of the internal pressure of the quantum membrane: current is leaving the pyramid.

- The release opens room for the quantum pressure from below again.
- As result, the impact of the pyramid on the external magnetic field, meaning the release of the anti-electron process surplus becomes limited again.
- The *Copper* process on the top works like a pump: sipping up the conflict from the pyramid below.

As the experiment proves: the potential of the pyramid in the first seconds *1.5V*, after stabilisation *0.739V*. The wires built into the other pyramids are not so sensitive to the impacts, but the voltage of the stabilisation is of higher value, around *0.8-0.9V*.

With the release of the electricity by one of more pyramids, their internal pressure is going down. The release is less and the intensity of the external magnetic field is decreasing as well. The decreasing external magnetic field gives more room for the release of the internal pressure of the anti-electron processes, which results in the go down of the intensity of the generating electricity. This way the pyramids are communicating and the change in one pyramid has its effect on the others as well. The pictures below demonstrate this communication, measured during the experiment.

There are three pyramids separated from each other. The distance between them is around *10-20 m*. The surface of the *Cuprum* process cap on the top was connected with the soil.

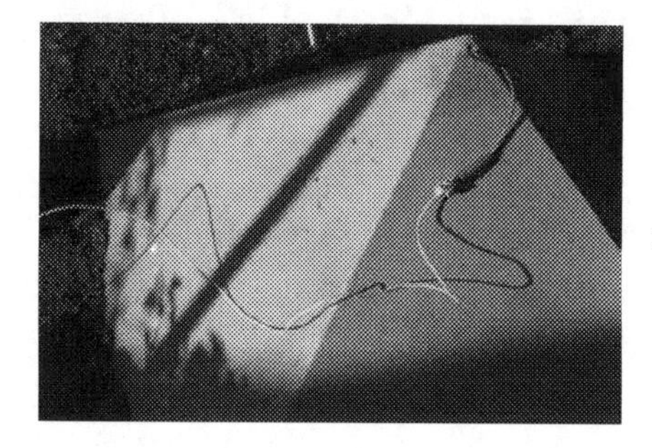

/a

The internal potentials of the two other pyramids were measured at the moment of the connection until the stabilisation of the voltage and the current.

/b

Picture 8.7

Pic. 8.7

There is only one difference between the other two pyramids: the number of the wires within one of them is much-much more (the digital meter on the left) than in the other, with three wires only.

Picture 8.8/a
voltages before the connection

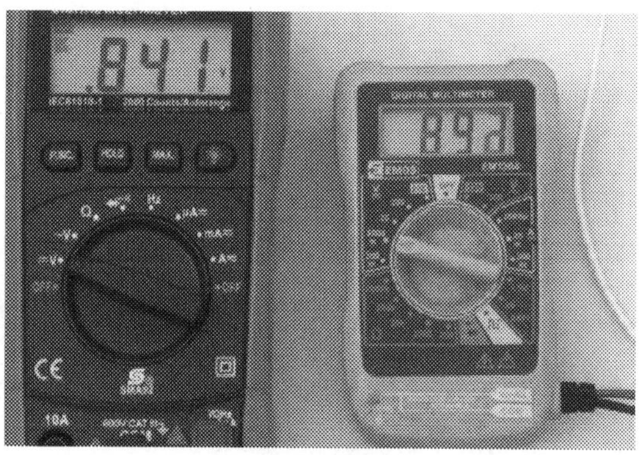

Picture 8.8/b
voltages during the connection

Pic. 8.8

The dynamism of the measured values illustrates the difference: The drop in the voltage in the pyramid with more wires is less: *8 mV* only, while the drop in the other is *18-19 mV*. The relation in the values of the currents is obviously the opposite: the amperage in the pyramid with more wires is much higher before and the drop is *20 μA*, while there is no measured drop in fact in the other.

Pic. 8.9

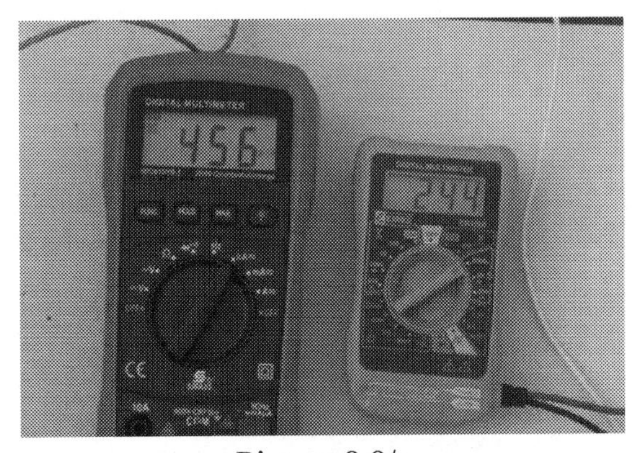

Picture 8.9/a — the values of the currents before the connection

Picture 8.9/b — the values of the currents during the connection

The experiment proves that this is not really about the compensation of the "loss" in the pyramid with the *Copper* process on the top by the other two pyramids, rather the accommodation of the other two to the new external conditions, to the less strength value of the external magnetic field.

The earlier findings, with reference to the Pictures 8.1 and 8.2, show that the conflict and the internal temperature and heat generation deep inside the pyramid are higher. The surfaces and the edges of the pyramids are more exposed to the external conditions. The deeper the conflict is, the higher is the temperature. The higher the internal conflict is the higher is the efficiency of the pyramid.

The only way for having a pyramid impact, the incoming and the released intensities of the anti-electron process impacts from the *Earth* shall be of different dynamism.

8C1 The unit value of the incoming intensity for the entire height of the propagation shall be: $\frac{di}{dS_{in}} > 1;$

8C2 while the unit value of the released intensity for the entire length shall be: $\frac{di}{dS_{out}} < 1;$

8C3 The unit value is the relation of the absolute value of the intensity to the area of the surface: $i_u = \frac{i}{S};$

The relation usually is: $i_u = \frac{i}{a^2};$

The two, opposite to each other tendencies of the change with reference to 8B1 have their most efficient integrated relation.

The unit values of the intensities of the quantum impacts of gravitation are equal everywhere. This also means that the external pressures the quantum membranes of the

elementary processes meet are permanent and of the same unit values in each case, whatever the elementary compositions of the materials of the pyramids are.
In the case of the elementary processes with less density, there is more room for the increase of the pressure of the quantum membrane. But the quantum impacts of the gravity from below cannot overwrite the standard density of the elementary process (without damaging the elementary process itself). The surplus is released, influencing the intensity of the external magnetic field, enabling with that the communication of the pyramids.
The unit values of the communication are equal; the differences in the absolute values however might be significant and unique and depend on the size of the pyramid.

The dynamism of the internal conflicts determines the direction of the communication.
The internal demand of the increase of the internal pressure of the quantum membrane in one pyramid results in the decrease of its external impact and allows this way for the others for increasing their impact. The increasing pressure initiates the generation of the *blue shift* surplus of the electron processes in parallel. The increasing external impact brings down the electricity generation.

8.3.1 Subjects and other elementary processes placed into the pyramid

S. 8.3.1

The materials of the pyramids, with their close to the balance status limit their impact on the external magnetic field. Placing other materials into the pyramid makes the impact wider and more efficient. There is more room for the change of the pressure of the quantum membrane with elementary processes of less density placed the inside of the pyramid. The structure remains the same: the envelop walls with high density transport and strengthen the impact on the external magnetic field, communicate the same way.

The pyramids have strong healing impact as well. The elementary processes of the living beings placed into the pyramids are driven for coming back to their balanced status. The intensity of the collapse of the neutron processes will be raised until reaching the standard density and this way balanced status.

8.4
The internal impact and the quantum communication of pyramids

S. 8.4

There is a solenoid with air core placed into the centre of one of the pyramids at the height of 1/3 from the base.

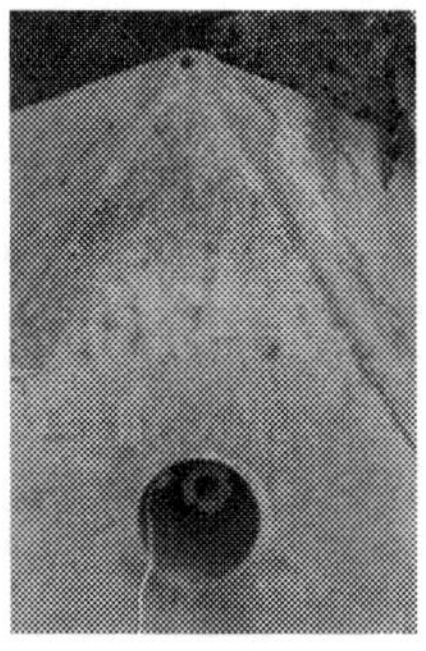
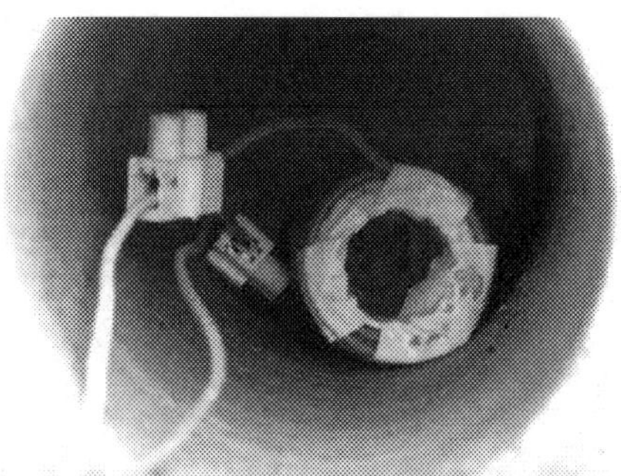

Picture 8.10/a/b/c

Pic. 8.10

This is the area with the most efficient and active internal conflict. The electricity source of the supply of the solenoid is an accumulator with 12V.

a/

Pic. 8.11

Picture 8.11

In parallel with the magnetic impact inside the internal potential of the other pyramid was measured.

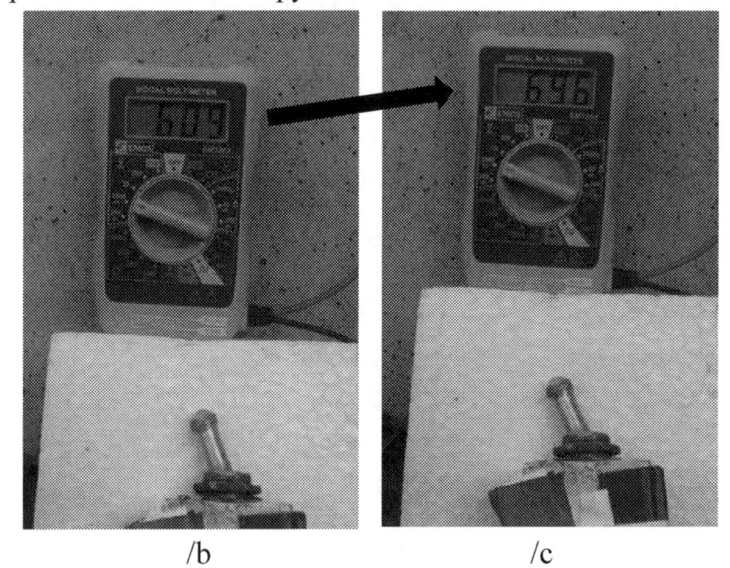

/b /c

The position of the switch means the solenoid in the other pyramid is on

The potential in the other pyramid grows up from *609 mV* to *696 mV*.

During the increase of the potential the value of the current went up from *89 μA* to *113 μA*.

Pic. 8.12

Picture 8.12

This experiment further proves the quantum communication of pyramids: the direct change of the internal magnetic field of the pyramid is influencing the energy potential of the other pyramid.

Communication works in both ways:

1. the release of the energy from the pyramids results in the change of the energy potential of the other pyramids;
2. the change of the internal potential of the pyramids has its impact on the release of the intensity potential of the pyramids.

The temperature of the solenoid, loaded inside, is also growing. The increasing heat inside has also its impact on the internal potential. While the initiation obviously is coming from the solenoid, it is difficult decide which impact is the more influencing one.

Picture 8.13/a

Picture 8.13/b

As the pictures demonstrate it, the internal temperature within the pyramid with the solenoid was 63.8°C, the external temperature and the temperature of the external surface were 9°C and 9.7°C.

Pic. 8.13

S. 8.5

8.5
The measured resistance is the proof of the external magnetic field

The magnetic impact of the *Earth* is the initiation and the magnetic impact of the pyramids is the realisation of the function, as it were discussed earlier.

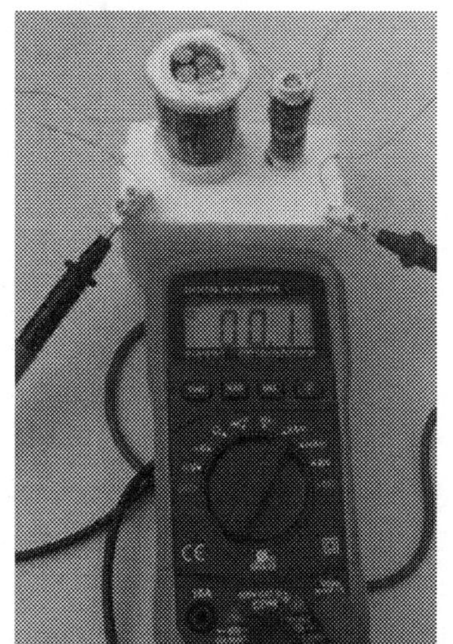

There is a picture of small scale electromagnets on the left; solenoids spooled with coil-wire, size of 0.118 mm.
The electro-meter is connected directly to the two solenoids.
There is no electricity supply within the circle, just the pyramid nearby.
The device shows $0.1\mu A$. In fact, the measured value was fluctuating between 0 and $0.1\mu A$ and stabilised on $0.1\mu A$.
There was no way and no need for measuring the value with more accuracy.

Picture 8.14

Pic. 8.14

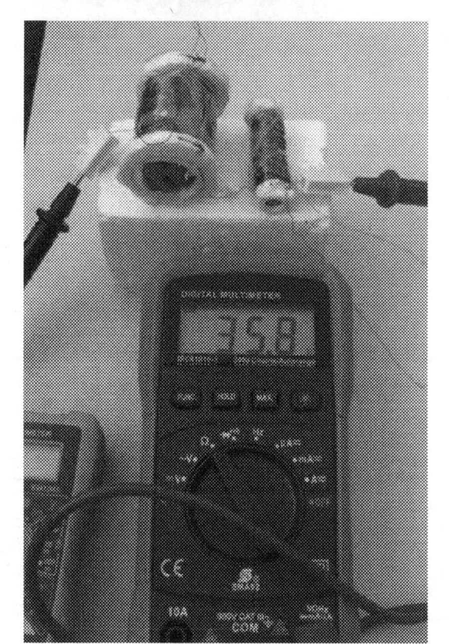

The value of the voltage was so small that it could not be measured. For the proof of the generation of the electricity it was important to measure the resistance of the circle.
And the measured resistance was *35.8 Ohm.*

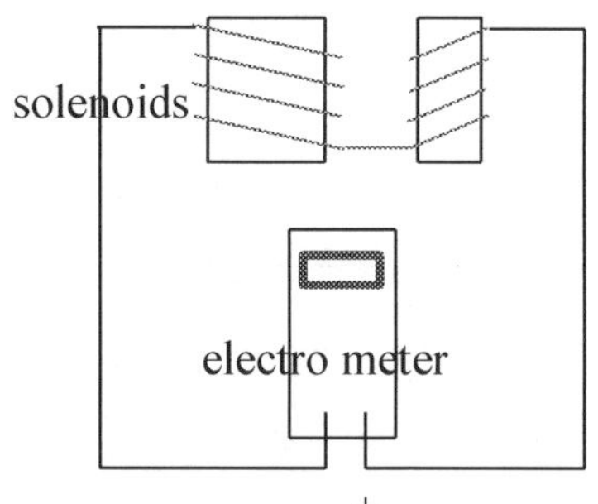

Picture 8.15

Pic. 8.15

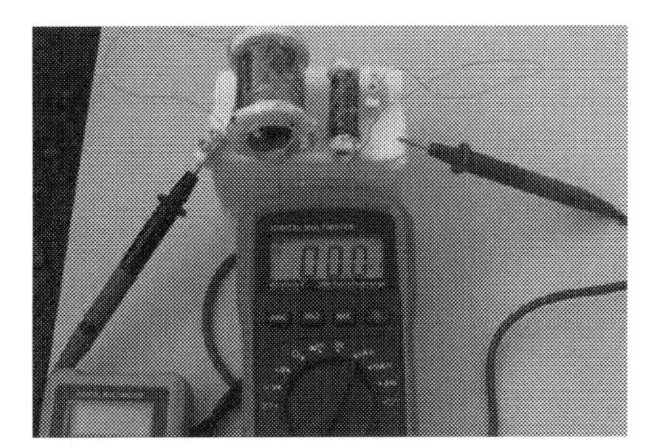

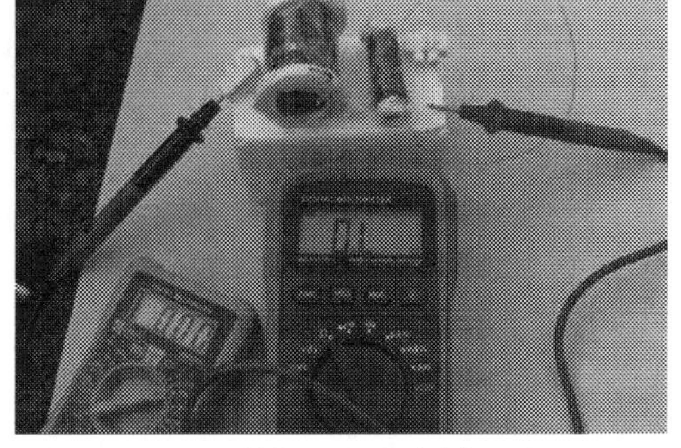

Pic. 8.16

Picture 8.16/a Picture 8.16/b

8D1 The no connection as above means the current is zero and the resistance has no meaning.

The resistance is a relation: $R = \frac{U}{I}$; The conductance is $G = \frac{I}{U}$;

The resistance and the conductance are expressing the relation of the potential and the current of the circuit. Therefore the values of the voltage and the amperage cannot be zero. In the case of $I = 0.1\ \mu A$; and $R = 35.8\ Ohm$, the voltage shall be: $U = 3.58\ \mu V$

Without connection: $I_{pir} = 0$; and $R = 35.8\ Ohm$

With connection: $I_{pir} = 174\ \mu A$; and $R = 35.7\ Ohm$

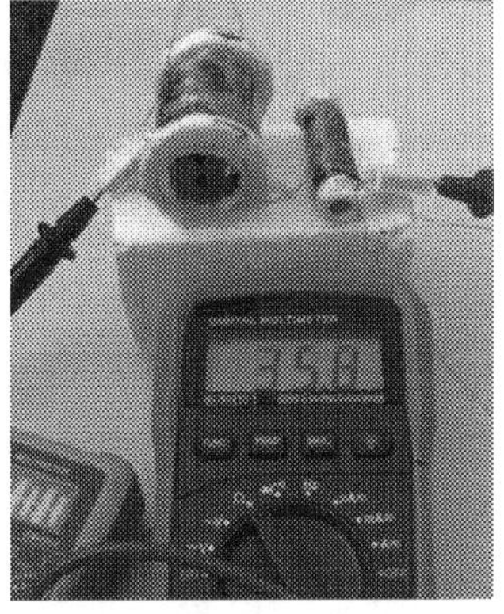

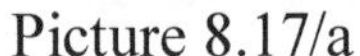

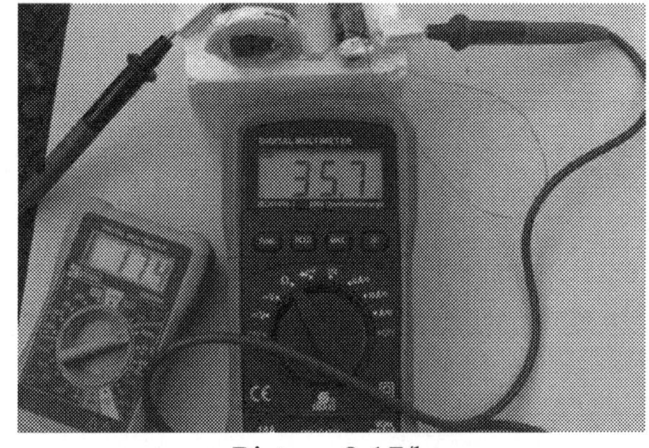

Pic. 8.17

Picture 8.17/a Picture 8.17/b

There were only slight changes in the measured values of the resistance, when the *Copper* process cap on the top was connected with the soil and releasing *174 μA*.

The solenoids with air core generate not measurable electricity at not measurable potential. But the energy and the magnetic field is here, surrounding us!

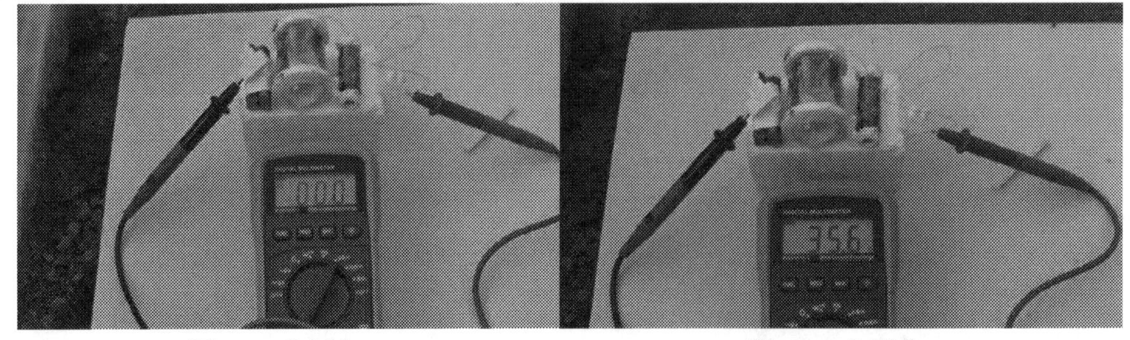

Pic. 8.18

Picture 8.18/a Picture 8.18/b

The values of the resistance are the only information about the magnetic field around us. The resistance in this case however represents parameters, diapason of micro values. And these values can be measured not just close to the pyramid but at longer distances from the pyramids as well. In fact this resistance value can be measured everywhere.

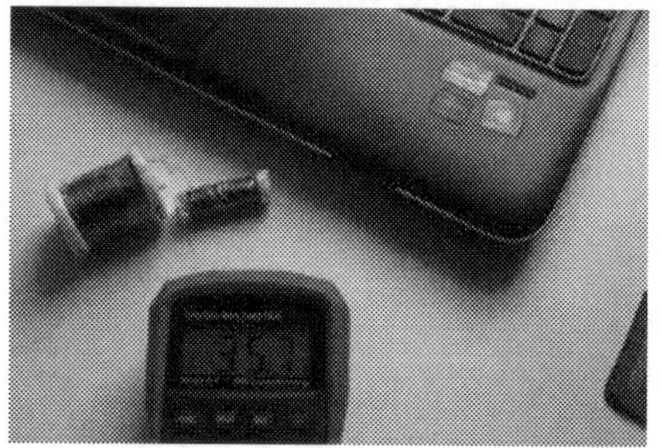

In a room near to the computer, in the car at far distance, in another town as well.

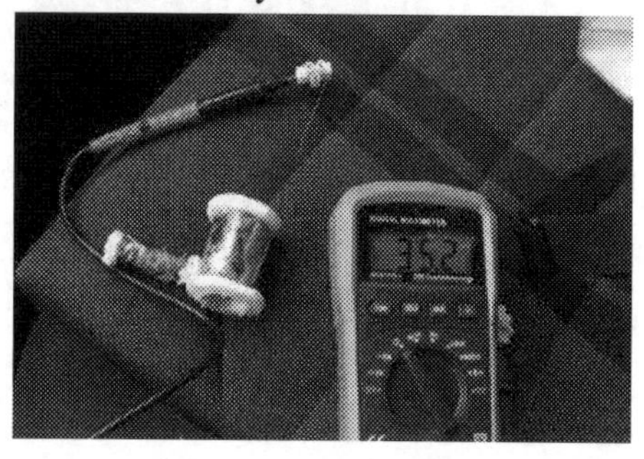

/a Picture 8.19 /b

Pic. 8.19

This might characterise the overall acting magnetic field on the *Earth* surface, which can easily be impacted by the existing pyramids.

The findings prove repeatedly the two equal sides of the events: if the electricity generates magnetic field than the magnetic field is also capable for the generation of electricity.

8.6 The practical benefit of the pyramids

S. 8.6

The communication, the computer and the internet technologies are the basics of our civil and business lives today. Our space-time on the surface of the *Earth* is not just full with the quantum impacts of the information, but also loads it up with intensity, energy and heat. The permanent conflicts of the intensities are acting energy potentials, which mean real risk for causing the greenhouse effect, the global warming and the climate change.

The system of the communication, built up on the pyramids may mean an installed and controlled total capacity of certain value and this way not just will limit the load but also harmonise it with the use.

The best way for the comparison of the energy intensities inside and outside of the pyramids is the comparison of the densities of the elementary processes of the communication.

$\frac{dmc_p^2}{dt_i\varepsilon_p}\left(1-\sqrt{1-\frac{(c_p-i_p)^2}{c_p^2}}\right)=n\frac{dmc_a^2}{dt_i\varepsilon_a}\left(1-\sqrt{1-\frac{(c_a-i_a)^2}{c_a^2}}\right);$	where n means the relation of the densities

8E1

$c_p > c_a$ and $\varepsilon_p < \varepsilon_a$ and $n \gg 1$, are the main reasons of the intensity difference and the generating conflict.

The intensities of the elementary processes of the air and the granite are almost the same. The difference in the atomic weights is also not significant. The massive and decisive difference is in their densities.

The average density of the *granite* (the pyramid) is $\rho_g = \rho_p = 2.65 - 2.75$ g/cm^3
The density of the air (oxygen) is $\rho_o = \rho_a = 1.429$ g/dm.
The relation is: $n = 1812$
$\varepsilon_a = \varepsilon_O = 1.0015$; and $c_a = 299670$ km/s;
$\varepsilon_p = \varepsilon_a = 1.000$; and $c_p = c_g = 299973$ km/s, as being in balance status on the *Earth.*
The relation of the quantum drives is: 1.0035 for the benefit of the pyramid.
But the acting quantum drive of the elementary processes inside the pyramid is $n = 1812$ times more even in normal circumstances (without any conflict) than the quantum drive of the air (oxygen) outside.

9
The free energy

The accelerator of the *Hydrogen* process generates free energy.

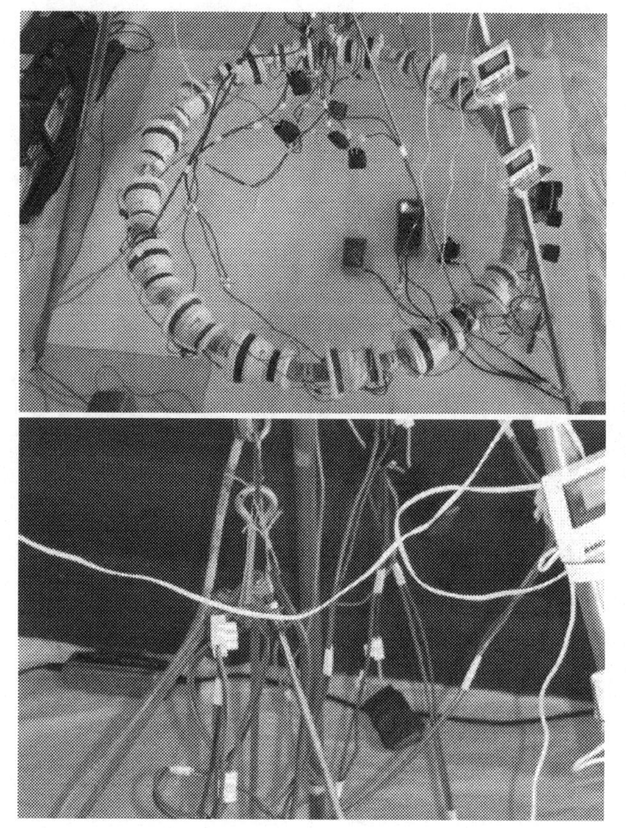

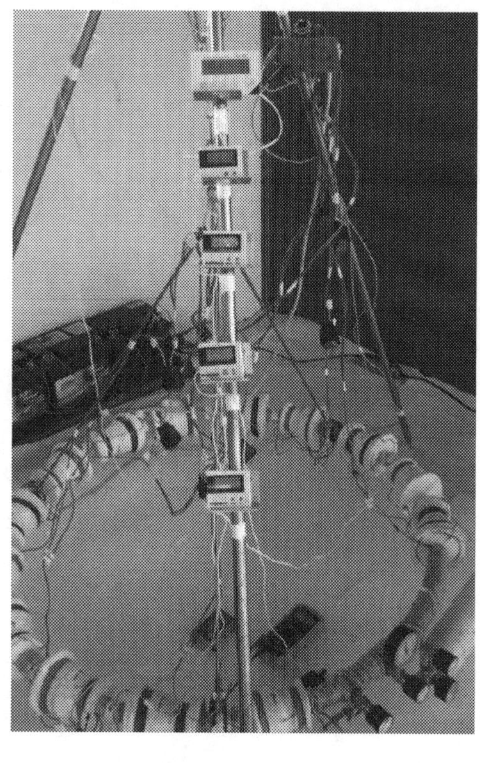

Picture 9.1. a/b/c

Pic. 9.1

There is a circular channel – with the *Hydrogen* process (2 litre at 5 bar) inside with 14 electromagnet quantum drives all around – suspended up on a support with three legs. The weight is measured by a balance, fixed to the top of the support. There is also a frame for lowering down the accelerator on and lifting it up from the balance.

Picture 9.2/a/b

Pic. 9.2

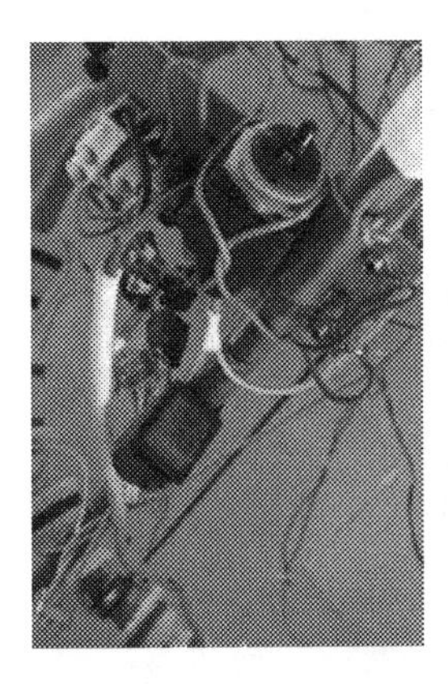

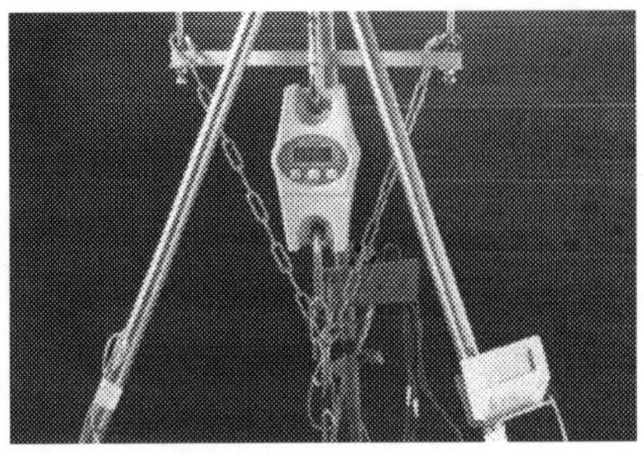

the lowering and the lifting frame

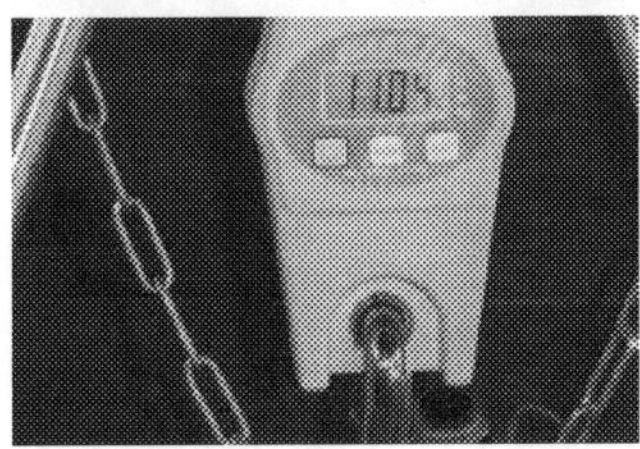

Pic. 9.3

Pictures 9.3/a/b/

The circular channel is for the acceleration, the balance is for measuring the weight and the weight difference, result of the acceleration. The permanent use of the balance was only possible if the measures were made in the certain sequence of the lowering down and the lift up of the accelerating channel. The range of the measurement of the balance is 20 kg. While the precision of the measurement is only 0.5%, the reproducibility of the balance is 10 g. The exact weight values were not important. The focus was on the difference of the measured values at the two ends of the down and the up directions. The mass of the load (the accelerating channel with all necessary devices) to be measured was each time always exactly the same. The measured weight values were compared with each other at the ends of the load and the lift. There were differences between these two values. The difference is generated by the conflict of the *quantum impacts* of the *Hydrogen* process and gravitation.

- The lowering down means loading the balance by the channel with the *Hydrogen* process, from zero to the full weight of the channel.
- The lift up means the take off the full mass of the channel from the balance step by step up to reaching the zero weight, taking over the full weight by the loading frame again.

The weights, measured at the end of the load always exceeded the weight measured at the end of the lift up. There were made more than 200 measures!! The difference usually is exceeding the 10 g reproducibility limit. There were only a couple cases with measured equal values.

There was obviously no way for measuring the precise weight values, having voltage of $U = 37 - 12$ V, and current above $I \geq 12$ A and with *Hydrogen* process of 1 litre and $p = 5$ bar. The objective of the experiment was only the control of the impact of the acceleration and finding the tendencies of the change if any.

Ref. S.3.6

There were electromagnet quantum drives all around the channel for speeding up the *Hydrogen* process. The drives, with reference to Section 3.6, are based on the attracting and the repelling impacts of the electromagnets. The increasing intensity and the increasing temperature of the *Hydrogen* process generate conflict with the quantum impact of *gravitation*. The inrush current of the 14 electromagnets of the start was high, the mechanical impact of the acceleration with loading-lifting sequence was measured.

The experiment might be followed on the YouTube address:
https://www.youtube.com/watch?v=9h94PCnvD-c&t=229s

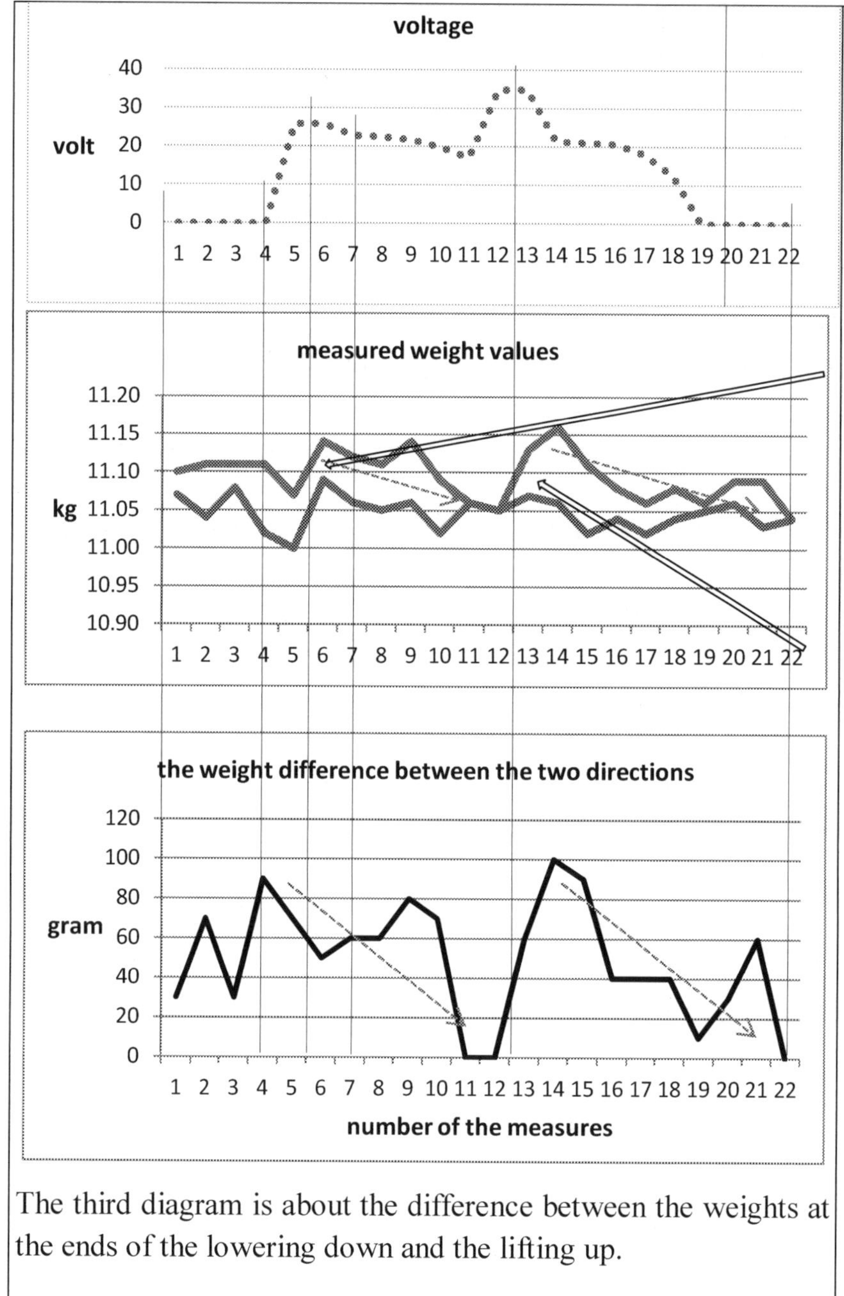

The third diagram is about the difference between the weights at the ends of the lowering down and the lifting up.

Diagram 9.1/a/b/c

The acceleration starts with U = 25.9 V. The inrush current is so high in the first moments of the load that it results in attractions on both sides. In fact, the electromagnets hold up the motion: the mass is increasing.

The same increase of the mass happens in step 13, at the time of the connection of the third accumulator: voltage is up to 34.7 V.

Weighs are decreasing in between. The increase in the upper direction is slight (the line below), in the downward direction (above) is more significant.

The difference is the proof of the lifting force of the conflict.

Diag. 9.1

The conflict between the increasing quantum impact of the *Hydrogen* process, as the result of the acceleration and the quantum impact of the gravity is easing the weight in both directions; during the load and the lift as well. The conflict has its push-up impact in both cases. The 9.1/b diagram also shows that the easing impact of the conflict is more significant on the direction of the load downwards. While the lifting force is obviously more, the difference in fact is determined by the load as:

The difference between the "real weight" and the weight of the load is

9A1 $G - \alpha G = G_d = G(1 - \alpha)$

The difference between the "real weight" and the weight of the taken off is:

9A2 $G_u = G - (G_d + \alpha G_d) = G - (G - \alpha G) + \alpha(G - \alpha G) = G - (G - \alpha G)(1 + \alpha);$

9A3 $= 2\alpha G - \alpha^2 G = G(2\alpha - \alpha^2) = 2\alpha G(1 - \alpha);$

9A4 If $\alpha \ll 1, G_d > G_u;$ meaning: the downward direction is the dominant in the easing of the weight

9A5 if $\alpha > \frac{1}{2}, G_u > G_d;$ meaning: the upward direction is the dominant in the easing of the weight

The electromagnet quantum drives are designed for the normal load. The inrush current is blocking the flow; the electromagnets in these cases generate high quantum impacts themselves. It increases the weights and increases the gap between the two directions.

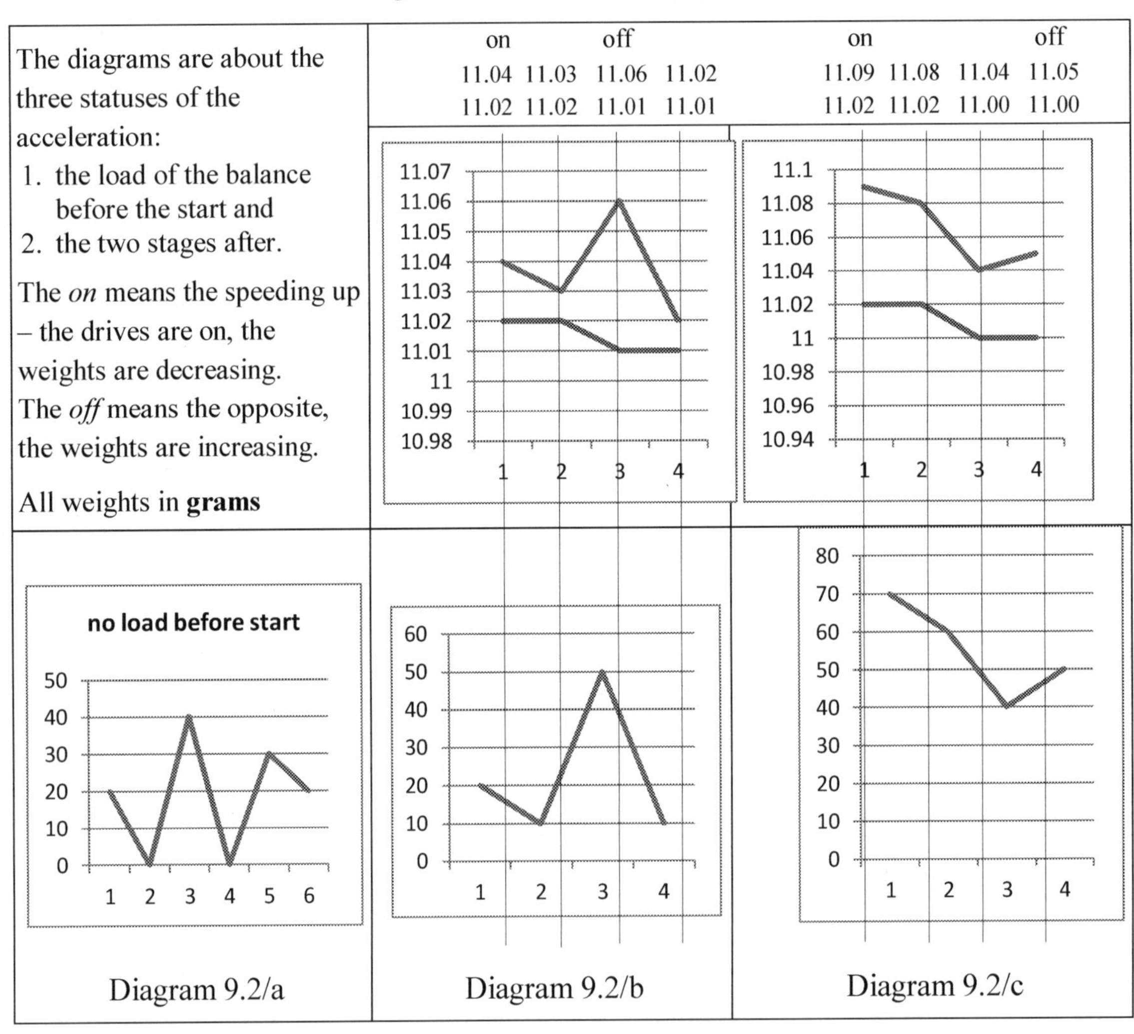

Diag. 9.2

Diagram 9.2/a Diagram 9.2/b Diagram 9.2/c

There are two parts in the /b and /c sections of the diagram above. The parts with two lines show the measured weights in the directions down (upper) and up (below). The single lines show the differences of the two. There were also measured differences even before the speeding up started. During the load however the differences were more significant and harmonised with the load of the electromagnets.

Diagram 9.3 is for the illustration of the relation of the weights measured on the downward and the upward directions.

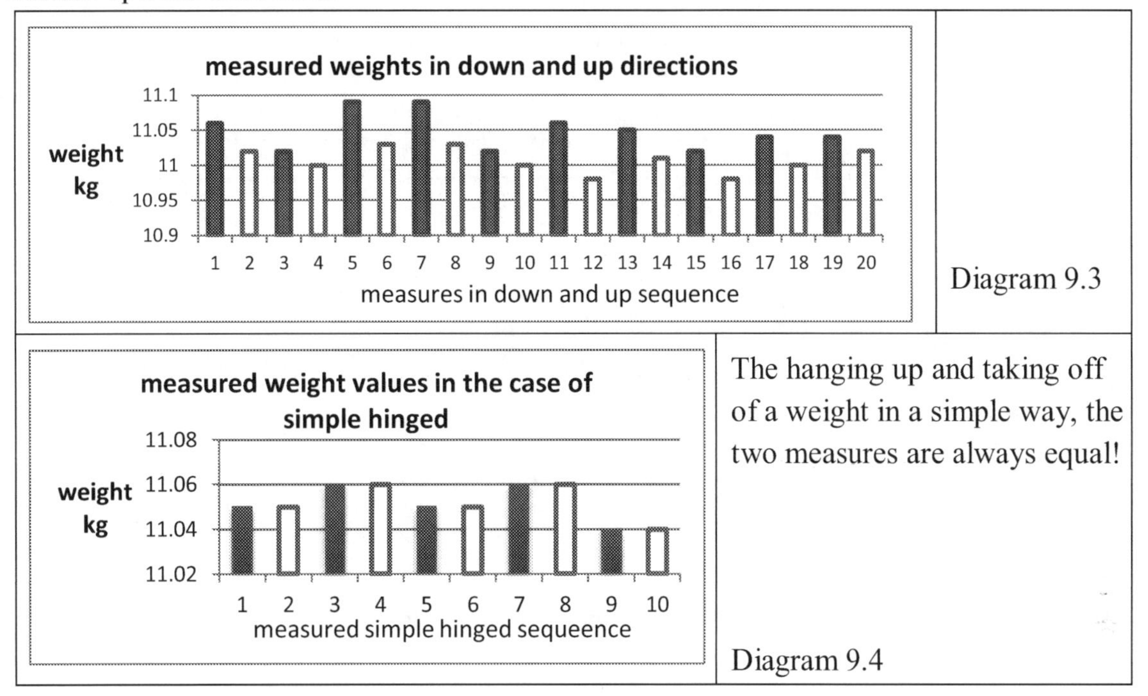

Diagram 9.3

Diag. 9.3

The hanging up and taking off of a weight in a simple way, the two measures are always equal!

Diagram 9.4

Diag. 9.4

There were also measures made using the average weight values $G = \frac{G_d + G_u}{2}$ of the down and the up directions. The results were similar. There are the phases of one and the same acceleration process on the Diagrams 9.5/a/b/c/d with the load *on* and *off*.

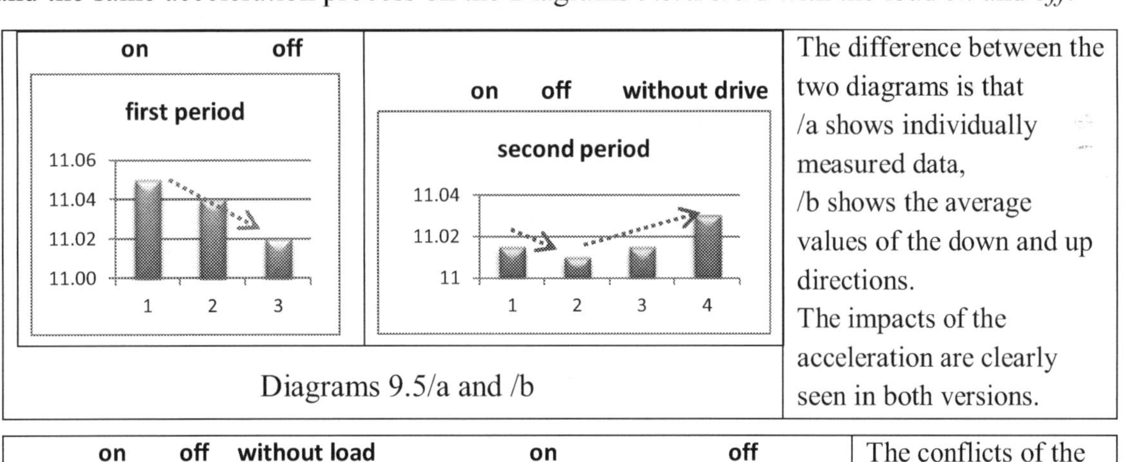

Diagrams 9.5/a and /b

The difference between the two diagrams is that /a shows individually measured data, /b shows the average values of the down and up directions. The impacts of the acceleration are clearly seen in both versions.

Diag. 9.5

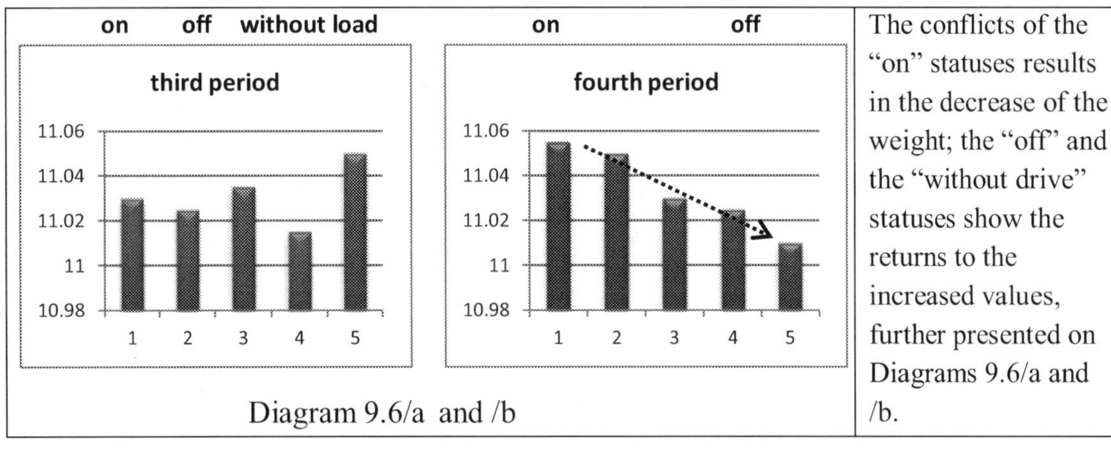

Diagram 9.6/a and /b

The conflicts of the "on" statuses results in the decrease of the weight; the "off" and the "without drive" statuses show the returns to the increased values, further presented on Diagrams 9.6/a and /b.

Diag. 9.6

The Diagram 9.7 below is for the illustration (separation) of the weight values of the still stand statuses from the load. The statuses of *on* and *off* are clearly seen, the control values of the still stand statuses are quite hectic. The switch off limit of the drive was 85-87 C^o .

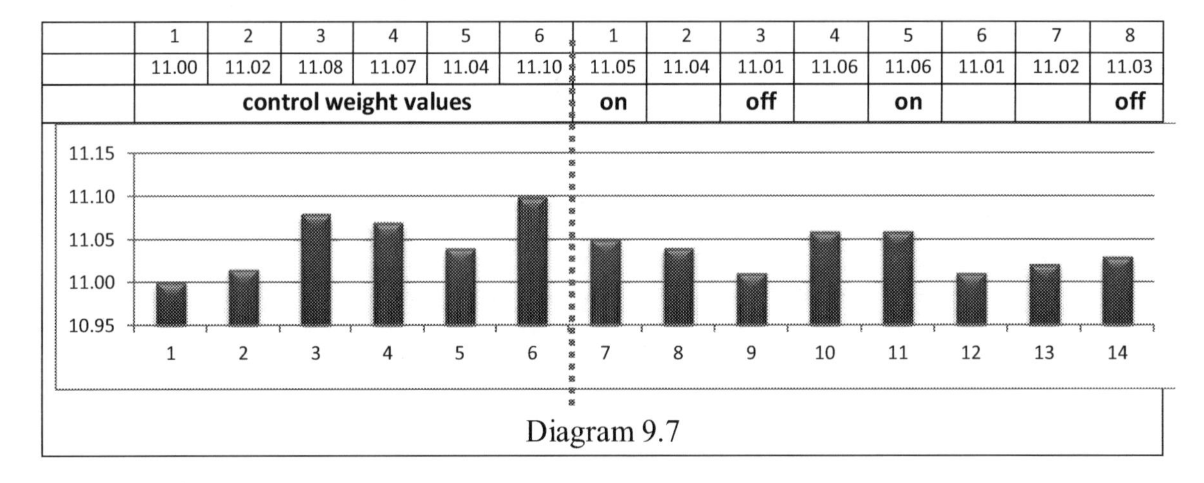

	1	2	3	4	5	6	1	2	3	4	5	6	7	8
	11.00	11.02	11.08	11.07	11.04	11.10	11.05	11.04	11.01	11.06	11.06	11.01	11.02	11.03
	control weight values						**on**		**off**		**on**			**off**

Diag. 9.7

Diagram 9.7

S. 9.1

9.1
The generation of free energy

The lift up effect of the conflict means the decrease of the weight of the channel with the acceleration of the *Hydrogen* process. The decreasing value of the weight is proportional to the conflict with the quantum impact of the gravitation.

The intensity (energy) of the drive of the electromagnet is: $P_e = U \cdot I$ (9B1)	The loss of the intensity (energy) of the weight is: $P_m = \frac{dW}{dt};$ (9B2)
If the intensities of the drive and the loss are equal, the balance is: (9B3)	$U \cdot I = \frac{dW}{dt};$ and $(U \cdot I)dt = dW;$

9B4 Meaning: $(U \cdot I) = \xi_M \frac{d}{dt}\left(\frac{Fds}{dt}\right);$	where coefficient ξ_M marks the efficiency of the mechanical work.
9B5 The mass of the channel does not change. The process is not about the transformation mass into energy, therefor $G = F = const$ formula shall be written as	$(U \cdot I) = \xi_M F \frac{dv}{dt};$
9B6 For having a lift, the force of the conflict shall be more than the weight of the mass to be lifted: $F > G$	$(U \cdot I) > \xi_M \frac{d}{dt}\left(\frac{Gds}{dt}\right);$

(In the case $F \leq G$ there is no lift; in the case $F > G$ lifting process does not stop.)

9B7 The intensity of the lifting process results in height s. It means the intensity of the impact can be expressed $\frac{ds}{dt} = v.$ This means in fact going out from the conflict with speed v.	$\frac{\Delta s}{\Delta t} = \frac{ds}{dt} = v.$

The decreasing weight, with reference to the 9B6 increases the height of the lift up:

once $G_2 < G_1 \rightarrow s_2 > s_1$.

If the move of the equipment upward is free, without holding it back, the conflict disappears, it is released.

In the case the lift up is hold back – for example the channel of the acceleration is fixed below the *Earth* surface – the only difference is that the release of the free lifting up is missing. The conflict between the *Hydrogen* process and the gravitation is generating the same way. The conflict however in this case generates heat and the heat generation needs cooling: this way the acceleration of the *Hydrogen* process generates energy.

The value of the heat generation is equivalent to the supposed to be loss of the weight. There is obviously no loss in the mass of the equipment. Just the internal conflict generates more and more heat. In the case of permanent acceleration the conflict is escalating and the resulting heat generation in exponential.

The small scale experiment – with the accelerating channel, mass of *11 kg* and the *Hydrogen* process of 1 litre and 5 bar inside – well demonstrates the power of the conflict. It resulted in the decrease of the weight about *30-50 g* in average!!

Ref. 9B1- -9B7 9C1 9C2

With reference to 9B1-9B6 this is means the weight difference gives:

$\Delta(m_1 \cdot v_1) = X;\ Watt$

against the speeding up intensity of the drive: $(U(V) \cdot xI(A) = E;\ Watt$

As the impact of the acceleration is escalating,

in the case of a driving force of *1000 MW* and the mass of the accelerator of 10,000 kg and the easing of v = 0.0001 m/sec, the time of the reproduction of the energy, used for the speeding up is: $N = \frac{E}{x} = \frac{1,000,000}{10,000*60*60*24} = 11.5$ days, after the acceleration is generating free energy!

10
Space-time, information, continuity

There is not just the definition of the space is missing from the academic physics, but also the real definition of the time as well. What the time is about?

The answer can be given easily if the explanation is based on processes.
Each process has its certain intensity. The intensity of the process defines the time duration while it happens. The reciprocal value of the intensity is the phenomenon we call *time*. It directly means there is no time without process, which is valid in its reverse format as well: there is no process without time. The definition is difficult on the basis of particles. The inclusion of the lifetime into the definition of the particles strongly needs process based explanation.
There is here also a principal point:
All processes have their typical end stage, where the very-very last-last portion of the event shall remain being not completed.

- all events, with internal driving force have that very last status, when the capacity, the intensity (the internal energy) is just not enough for the completion of the process; there is no way an event could cancel itself;
- the completion with turning into "nothing" is not just impossible; because "the nothing" as such cannot exist, but there would be no way to restart from the "nothing" status;
- the very last information of the matter remains acting as quantum impulse forever.

The intensity obviously can expire in the communication or in conflicts with other processes. But the completion as such for separate events, processes with self-drives is simply impossible.
The events or processes provide us not just the definition of the time, but also the definition of space as well. The remaining and forever acting last portions of the events, the quantum impulses of the processes shall accumulate somewhere all around. They are the ones making up the space! The more the events are, the larger is the space.

Events – time – space. Events happen in space-time. There is no space and no time without events. There is no event without space and time. The space-time in fact is about events.
The intensities of the events vary. Each event has its intensity and time count, and this way its time system. The time systems vary and they might be compared. The durations of the events within their space-time may differ from ours on the *Earth* surface and therefore while the events are one and the same their measured duration might be different.

The time systems and the time counts are relativistic.

If we state that the estimated lifetime of the proton process is as many-many years on the many-many power in our time system, it easily might be the duration of a wink within the own time system of the proton process itself. This might also be valid for the particles with extremely short lifetime, just with the opposite way. There is no difference. The principle is one and the same. Space-times exist in parallel!

The universe means the complexity of processes, quantum impacts and signals, the unity of time and space, the accumulating quantum impulses. The processes carry information about the time and space, about the communication and the conflicts of the elementary processes with different intensities and speed values. Matter is the complexity of the information with the infinite variety of intensities, space-times and volumes.

And there are still valid concerns related to the official academic physics.

The gravitation as the curvature of the space will always generate concerns until the definition of the space is missing.
The duality of the photon itself is the best example of the conflict and the contradiction of the existing views about the universe. As it supposed to be a *particle*, it represents the view of the universe with an end. As supposed to be a *process* with intensities (waves) it means a universe without end.
Is the light the phenomenon of the duality of the universe? Or is it the quantum signal/impact being handed over and over from quantum impulses to quantum impulses, representing this way the endlessness of the universe?

What happens at the moment when the source of the light is switched off?
Does the last photon, which is released at the moment of the switch off flies through the distance to the lightened subject? The fly needs not just drive, but also space with definition. Has each photon a separate drive; especially the last one? What does happen with those photons, not reaching the lightened subject? Does the photon disappear at certain distance? What is the status of its disappearance, since they also cannot turn into nothing?
There is no way for giving the valid answer in the conventional way of the physics.

Based on processes, the propagation of the light signal immediately stops at the moment of the switch off all along of the trajectory. The quantum impulses of the space stop receiving and handing over the signal. There is no light signal anymore for handing over to each other. Once the source is off, the propagation stops immediately.

The answers on these questions might be given with full confidence only if the light signal is considered as propagating quantum impact within the quantum space!
This approach gives the correct definition of the speed of the light as well. The speed of the light is the speed of the quantum communication, the speed of the quantum signal, propagating, handing over from quantum impulse to quantum impulse of the space.

The speed of the communication depends on the intensity, the energy of the source of the signal carrying the information. The energy of the source drives the signal ahead. Any obstacle on the way of the propagation increases the loss. The loss is either becomes recovered by the energy from the source or the signal disappears. The time count (the duration) of the propagation of the information with high intensity is short, with low intensities is long. Events happen within their own space-time with certain intensities. The space is one and the same for all space-times and for any kind of processes, meanwhile the systems and the processes themselves have their own intensities and time counts. The formulation of the value of the speed of the quantum communication in the unified and common space *depends* on the time count, the intensity of the space-time system.

There is no information/signal/matter without an event, without space and without time. There is no basic speed value of the quantum communication as such within the universe. The speed of the quantum communication depends on the intensities of the space-times; the time count of the system. Therefore the free variety of the space-times result in the free variety of the formulation of the speed values of the quantum communication. The value of the speed of light depends on the space-time of the measurement. It has its certain unique value on the *Earth* surface.

The speed of the quantum communication and the intensities of the space-times vary; the intensities of the magnetic impacts of the space-times are equal; the strengths of the magnetic fields depend on the number (the intensity) of the acting magnetic impacts. With the increase of the values of the quantum speed and the intensities of the space-times the strength of the magnetic field is also increasing.

The quantum impulses are generating as entropy products in the electron and the anti-electron processes and establishing the space/universe.

There is no way the *neutron* and the *anti-proton* processes will start if the electron and the anti-electron processes, the drives of the collapse run out to zero. The proton and the anti-neutron process have their internal drives from the accumulated energies of the collapse to start from the inflexion, the status of the full intensity.
There is no matter with zero intensity, the "nothing" as such does not exist. The collapse needs not just external drive, but also a starting intensity. The nothing as such cannot be driven.

The sources of the energy of the expansion of the proton, the electron, the anti-neutron and the anti-electron processes are the internally accumulated intensities of the elementary process. The collapse of the neutron and the anti-proton processes are driven by the release of this energy, but these processes at the same time are accumulating the intensities and re-establishing the source.

The turn from the self-expansion to the collapse to be driven represents the elementary balance within the elementary process

The expansion starts from the inflexion of the anti-proton/proton processes: e_{pinfl}; 10A1

The value of the intensity decrease of the self-driven expansion process (the proton and electron processes): $\Delta e_p = e_{pinf} - e_{qi}$; 10A2

There is internal elementary communication between the neutron collapse driven by the electron process and the proton process providing the energy of the process:

$\Delta e_n = e_{ninf} - e_{qi}$; 10A3

The utilised intensity as the electron process drive is: $\Delta e_{drive} = e_e - e_{qi}$; 10A4

The drive potential, which cannot be used: $\lim \Delta e_{qi} = \lim(e_{qi} - 0) = 0$; 10A5

The balance is: $e_{pinf} = e_{ninfl} + e_{qi}$ 10A6

Δe_{qi} , the quantum impulse is free, accumulating and establishing the space! 10A7

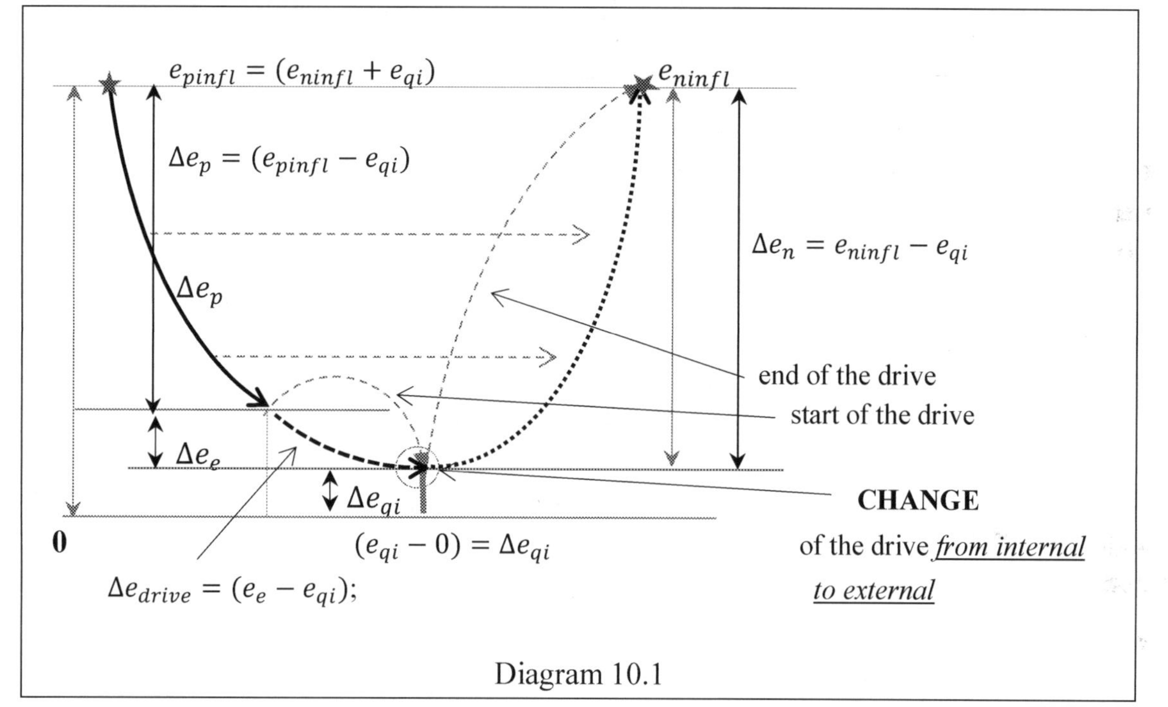

Diagram 10.1

Diag. 10.1

The capacity of the proton process, at the anti-proton/proton inflexion contains the intensity of the electron process as well:	$\frac{dmc_x^2}{dt_p}\left(1 - \sqrt{1 - \frac{v^2}{c_x^2}}\right)$;	10C1
	$\Delta t_p = 0 \div t_i$ and $v = 0 \div i$;	10C2

The electron process is the drive at permanent acceleration of $(c_x - i_x)$:	$\frac{dmc_x^2}{dt_i\varepsilon_e}\left(1 - \sqrt{1 - \frac{(c_x - i_x)^2}{c_x^2}}\right)$; and $\lim \Delta t_i = \infty$; and	10C3

until the last stage of: $\lim \frac{dmc_x^2}{dt_i\varepsilon_e}\left(1 - \sqrt{1 - \frac{(c_x - i_x)^2}{c_x^2}}\right) = 0$;	$\varepsilon_x = \frac{\Delta t_n}{\Delta t_p} = \sqrt{1 - \frac{(c_x - i_x)^2}{c_x^2}}$;	10C4

10C4 means, this is the remaining not utilised energy capacity of the electron process.

10C5	Neutron process,	$\frac{dmc_x^2}{dt_n}\sqrt{1-\frac{(c_x-i_x)^2}{c_x^2}}\left(1-\sqrt{1-\frac{v^2}{c_x^2}}\right);$
10C6	which is driven from $\lim\frac{dmc_x^2}{dt_i\varepsilon_e}\sqrt{1-\frac{(c_x-i_x)^2}{c_x^2}}=1$	and $t_n = t_i \div 0$ and $v = 0 \div i;$

10C6 means, the remaining energy capacity of the electron process is of infinite low value.

If the neutron process would contain the intensity remains of the electron process, the intensity of the quantum impulse, the neutron collapse, driven by the electron process would cover the intensity change of

10C7
$$\frac{dmc_x^2}{dt_n}\left(1-\sqrt{1-\frac{(c_x-i_x)^2}{c_x^2}}+\sqrt{1-\frac{(c_x-i_x)^2}{c_x^2}}\right)\left(1-\sqrt{1-\frac{v^2}{c_x^2}}\right)=\frac{dmc_x^2}{dt_n}\left(1-\sqrt{1-\frac{v^2}{c_x^2}}\right);$$

The equation here above 10C7 in this case gives a static balance. The neutron and the proton are the processes of the intensity exchange of the quantum communication. The collapse is taking over the energy/intensity of the expansion. But only
(1) if it is driven, as no collapse can drive itself; and
(2) the intensity of the start of the drive is not zero!

S. 10.1

10.1
The elementary processes are the carriers of the information but within the space-time

All elementary processes are processes with different intensities – in fact the information about the processes themselves. The intensity of the process determines the timeframe of the process, the speed of the quantum communication, the pressure of the elementary quantum membrane (the densities), all together specify the aggregate status and the lifetime of the information. The elementary cycles are acting in parallel.

The communication happens in space and time. The communication means messages, the quantum signals. The time defines the propagation, as event.

The messages appear in space and time. The space-times are the mediators and at the same time also the carriers of the information as well. The principle of the common space – built up from quantum impulses – for all the messages to be conveyed is: let it be! It is complemented by the intensity (the time) of the message: let it be as it wants to be.

The timeframe of the information defines the appearance of the message. Time systems exist in parallel and are embedded in each other. The space-times on the surface of the *Earth* and the space-time of the *Earth* as planet are the parts of the global system. While the time systems exist in parallel, the intensity identifies the space-time of the appearance of the event.

The space-times, the intensities of the information of the elementary processes between the *plasma* and the *Hydrogen* processes vary and are different. The infinite high intensity of the elementary conflict, the infinite high intensity of the inflexion is the *plasma* status itself. The *plasma* is communicating with the "external environment" through the elementary processes up to the *Hydrogen* process, the accumulation of all information, the initiation of the plasma-inflexion of the next cycle of the elementary evolution.

The life and the lifetime are not about the disappearance of the matter. It is about the completion of the function, but leaving its trace in the space, the quantum impulse, establishing the space itself for the continuity of the life and the new start!